T0280542

'A book published by seasoned practitioners and not just academics, worth a read because those who have written it have skin in the game.'
— **Prof Robert Whittaker,** *Director of Holdom Whittaker Adjunct at University of Canberra, University of Newcastle Western Sydney University, Australia*

Understanding Australian Construction Contractors

This book will provide emerging construction professionals with insights and information helpful for a successful career in the Australian construction industry. This work fills a critical gap and is written by two authors with decades of experience immersed in current issues. It provides a starting point for the next generation of Australian construction contractors.

Beginning with an overview of the industry, the chapters explore winning work, project operations, financial management, people skills and selling a successful business. The authors use case studies to enrich the content and include reviews and commentaries on some of the legendary management books. In addition, readers of the book will find answers to essential industry questions:

- Why is construction one of the best industries in Australia?
- What is its most significant conflict?
- Which are the three most consistently profitable sectors?
- What are the essential ten questions to answer for standardising practices?
- Is work acquisition more art or science?
- Is it a good idea to fire a client? Why?
- How to identify and address the office – field conflict?
- What is the job cost format for unifying project stakeholder information?
- What are the best key performance indicators for a construction contracting firm?
- What alignments are needed in general hiring and personnel management processes?
- What is the process in identifying and implementing a best practice?
- How do you value the market price for a construction firm?

This book should be read by anyone entering the built environment sector in Australia. Universities, Colleges and TAFEs can use this book in various construction business and operations management courses. Supporting materials are available through a website.

Matt Stevens has a significant background in the construction industry for over four decades and utilises this experience in research, teaching and engagement at Western Sydney University, Australia. His textbooks, *Managing a Construction Firm on Just 24 Hours a Day* and *The Construction MBA*, have been adopted by dozens of universities and associations worldwide. His research publications focus on Contractor Operations, Innovation, and Education. Stevens still advises construction contracting firms. In addition, he has coached student teams, recently winning Australia's Constructathon and reaching the finals of several international competitions.

John Smolders AM FAIB is a chartered builder with over forty years of construction experience. Currently a Senior Lecturer of Practice at Western Sydney University, Australia. An active member of the Australian Institute of Building (AIB) and the Master Builders Association (MBA), serving as Past President of the AIB and past Vice President of the MBA (Newcastle). John has formally represented the construction industry via advisory panels to government and universities, focusing on emerging constructors. John was awarded an AM (Member of the Order of Australia) for his significant service to the construction industry and higher education.

Understanding Australian Construction Contractors

A Guide for Emerging Professionals

Matt Stevens PhD and
John Smolders AM FAIB

LONDON AND NEW YORK

Cover image: © Getty Images

First published 2023
by Routledge
4 Park Square, Milton Park, Abingdon, Oxon OX14 4RN

and by Routledge
605 Third Avenue, New York, NY 10158

Routledge is an imprint of the Taylor & Francis Group, an informa business

© 2023 Matt Stevens and John Smolders

The right of Matt Stevens and John Smolders to be identified as authors of this work has been asserted in accordance with sections 77 and 78 of the Copyright, Designs and Patents Act 1988.

All rights reserved. No part of this book may be reprinted or reproduced or utilised in any form or by any electronic, mechanical, or other means, now known or hereafter invented, including photocopying and recording, or in any information storage or retrieval system, without permission in writing from the publishers.

Trademark notice: Product or corporate names may be trademarks or registered trademarks, and are used only for identification and explanation without intent to infringe.

British Library Cataloguing-in-Publication Data
A catalogue record for this book is available from the British Library

Library of Congress Cataloguing-in-Publication Data
A catalogue record has been requested for this book

ISBN: 978-1-032-26948-1 (hbk)
ISBN: 978-1-032-26947-4 (pbk)
ISBN: 978-1-003-29064-3 (ebk)

DOI: 10.1201/9781003290643

Typeset in Times New Roman
by MPS Limited, Dehradun

This book is dedicated to all Australian construction contracting professionals and their continuous improvement. My colleague, John Smolders has been the personification of this country's stellar built environment leaders-a sincere commitment to do things better each day and a willingness to share with young people.

Matt Stevens PhD

Working with Matt Stevens, the principal author of this book, has been an absolute pleasure. Constructors of any age will find this recourse very useful. Practical information contained herein has been designed to be easily digested and understood. A practical learning guide for the young and perhaps not so young constructor.

John Smolders AM FAIB

Contents

Acknowledgements

Many thanks to Dr Laura Almeida for her editing leadership of this book. The quality of the product is much improved.

Our appreciation goes to Western Sydney University and the School of Engineering, Design and Built Environment for their encouragement to explore construction contracting issues and connect with the industry in many ways.

Foreword

This book is one beginning point for you, the young Australian constructor, to grow your knowledge. Mushonga E. (2015) asserts it is critical, The success of societies depends, now more than ever, on the ability of young people to thrive amid relentless change". From our library of work, we want to share with you other insights from our years. This book is the product of over two years of writing and more than a collective seventy years of industry work. Perhaps there are many other books you have read. Each should be an attempt to give you many years of experience and observations in exchange for a short investment of time. To that end, we have detailed methods and ways of thinking to consider. Much of this is from a realistic strategy of leading and managing a construction contracting firm. This book is a snapshot of our collective research, observations, analysis and thinking when it is published. As stated, it is based on the reality of the construction contracting business in the built environment.

The purpose of this book is straightforward. We believe that helping instil a potential mental model" into the reader's thinking. All experts have mental models of what they should expect to see in a given situation. With this framework in a person's head, they can interpret a problem accurately and rapidly. Furthermore, assessments might be considered to be tests of the breadth and depth of the intended mental models. Learning outcomes could include these Models, which can be simple or complex; however, they are based on whether it be proactive such as planning or reactive such as crisis management. Linear Thinking" is what most of these intellectual frameworks are composed of (Senge 1990).

We sense that there needs to be a "backfilling" of knowledge. All generations suffer from a lack of foundation not passed along by the previous generation. Sometimes, it is never explained because that knowledge is assumed. That is the basics or first principles of how and why of many accepted construction practices. Sometimes, explaining them prompts young people to think of a better way.

Construction Management decision-making needs more leaders with robust mental models. Each good decision should pass through many discerning filters" and poor decisions would not pass through some of them, thus be rejected as a possible answer. These filters may be critical questions or graphical representations.

Transcendent mental models are where we start in this book. In construction management, we believe that some mental models are critical to discuss. For these select few, we attempt to represent them graphically. Butcher (2006) asserts that using diagrams with text improves a mental model than just text or a chart.

Our viewpoint is from the construction contractor's. We have structured this book from general to specific ideas. We apply them specifically to construction contracting operations. To simply state, this book's value is that it is a starting point for all for-profit construction organisations to contemplate. We have actively engaged more than 100 construction contractors in our careers as industry participants.

We work in academia with a good engagement focus on industry. We found both areas to be contributors to the construction industry. Indeed, our understanding from research and practice lends itself to our teaching. We try to stay on the edge between established and emerging applications to the construction contracting industry. This reminds me of a famous research joke, We know this finding works in practice; we are hopeful that it works in theory" (Isaacson 2014). More seriously, we hope our knowledge is beneficial to the industry to further your thinking.

We have found that many times, consultants like to sell complexity. Why the different approach? We know that simplicity is easier to teach in-depth and breadth. Everyone in the company can quickly adapt and apply this kind of input. As a result, companies are more productive quicker.

We think it is essential to state that we both have a heart of a teacher". We mean that the teachers care to inform others completely without consideration of a commercial transaction.

We hope that this book will raise thinking and conversation about construction transformation in Australia. We have personally seen this decline in other countries, including Australia (we have both worked there). Therefore, it is not a unique problem for Australia.

The solutions offered, we feel, will apply. So, this book is specific to Australia in its focus.

Construction companies face multiple problems when operating today. An executive's frustrations are real. The list is long, and many have emerged during our careers, such as politicisation, risk-sharing and human diversity. Some changes occur as new legal requirements and others as realities.

Older construction professionals know that success is relative and not absolute. Critical elements of a construction company's profitability, such as client relations, will never be perfect; technology is not bug-free", nor people are predictable. However, when issues such as these are better than your competition, you outpace them in safety, quality, or productivity.

Matt Stevens PhD with John Smolders AM FAIB
Sydney, Australia

Introduction

This book's rich depository of experience and research from industry professionals is a starting point for Young Australian Constructors. We have worked with over 100 firms in our combined experience and have condensed it in this work. It explains many of the current challenges and explores approaches to managing each well. We answer these questions and more:

1 Where does construction rank as a career in Australia?
2 What is the current state of construction multifactor productivity?
3 What is the nature of the conflict between productivity versus production, and how to overcome it?
4 Which are the three most profitable sectors consistently in contracting?
5 What is the first best practice in construction?
6 What are the starting ten questions to answer for standardising practices?
7 What is the effect on efficiency for 1% more practice adherence?
8 Is work acquisition more art or science?
9 What tendering measure provides a holistic view of effectiveness?
10 Is it a good idea to fire a client? Why?
11 How common is the office and field conflict, and what to do about it?
12 What is the single measure of financial performance for a construction contracting firm?
13 What tailoring is needed in general hiring and personnel management processes?
14 What are the hallmarks in identifying a best practice?
15 Should the implementation of improvements be rapid or steady? Why?

This book has dozens of graphics and tables to crystallise the concepts explained. Furthermore, the book's chapters cover:

1 The Australian construction industry
2 The dynamics of the construction contracting business
3 Work acquisition processes
4 Project operations
5 Financial management concepts and practices

6   Human resource management
7   Retiring from your construction contracting business
8   Case studies
9   Reviews of legendary management books and applying them to the business of construction

We believe that once you look inside the book, you will see its value for money.

1 The Australian Construction Industry

The Australian Construction Industry is multi-dimensional with many professional disciplines and market subsections in which people and companies operate. According to the Australian Bureau of Statistics (ABS), it is one of nine major industries. Its employment of more than 1 million persons and a company population of almost 400,000 firms make it a bell-weather for its economy.

Construction contracting organisations install all components that comprise shelter (such as commercial buildings and residential units), processing facilities (such as manufacturing plants, petrochemical installations and ports) and infrastructure (such as bridges, roads, wireless facilities and airports). Some of these companies, often called main contractors, organise new project installations with subcontractors and suppliers that furnish specialised products and installation services. Other companies repair and rehabilitate current structures for extended useful life. Still, other firms demolish the built environment when those man-made installations can no longer deliver their intended benefits.

Construction contracting employment positions may be categorised into five broad overlapping sectors: 1) Work Acquisition such as designers, estimators and tender managers. 2) Project Operations such as project managers, site supervisors and contract administrators 3) Financial Management such cost analysts and directors of finances 4) Business Operations, such as office managers and asset coordinators 5) Human Resources such as HR directors, benefits specialists and hiring coordinators. Each title is not universal to the construction organisation since the firm's size may require employees to be assigned two or more roles making for a novel title. Many companies exist that profit from these kinds of services to the construction firm as an outside contracted service. The range of exciting employment opportunities and career paths is exponential.

Overall Industry Data

The Australian Construction industry is a significant part of this country's economy. Manufacturing produces tangible products and makes for a fair comparison (Table 1.1).

As shown above, this industry is relatively efficient. Australian construction contractors are efficient. They compete in the third-largest economic sector by turnover and are the largest private employer. Being careful with risk and reward is a way of life in construction. The statistics comparing the construction industry

DOI: 10.1201/9781003290643-1

Table 1.1 Comparison of Australian Construction and Manufacturing Industries

FYE 2020	Construction	Manufacturing	Comments
Employment	1,104,000	831,000	Construction employment is 20% larger
Number of Businesses	394,496	86,226	Construction has four times the competitors
Average Entrants Annually 2016–2020	6.3%	1.9%	Construction has three times the new competitors than manufacturing
Average Exits Annually 2016–2020	14.0%	11.1%	More exits indicate more risk factors.
Turnover of Construction Work at Current Prices	$210,659,704,000	$405,091,000,000	Construction has approximately half the turnover.
Current Turnover per Company	$533,998	$4,698,014	Construction Turnover per company is 1/8th of manufacturing.
Value Added	$126,293,000	$107,479,000	Construction adds 59.9% value per turnover dollar whereas manufacturing contributes 26.4%

Source: Australian Bureau of Statistics.

with manufacturing – show that when basic comparative information is considered, the industry does much with little. This data is from the leading Australian data source, the ABS and tells a different narrative than outsiders have stated.

As a side note, a synergy exists between both industries as prefabrication and modularisation are emerging as a practical option for development, government and individual purchasers.

A significant part of the Australian built environment industry's robustness is facilitated by immigration from other parts of the world – both consumption and production. Generally, this trend has been dampened in the current COVID pandemic. However, our location close to the most populous countries in the world seems to guarantee this trend will rebound. In addition, there are other contributing factors, such as birth rates and rural-suburban-urban population movements, that increase construction demand. As history instructs, all elements are certain to change and being observant helps young constructors lead industry thinking.

One of the Best Industries in Australia

We believe that construction contracting offers much to many young people starting to think about industries and careers. We have concluded it is the best career a young person could choose. You may not believe it, but the facts will show that our business is unmatched. It offers participants long-term advantages. However, many people will argue. Let us share with you a dozen career benefits our industry possesses. You be the judge.

Constructors have a Cost Advantage in the World's Largest Asset Class

Once you understand the process in most construction professions and gain connections to build the most expensive personal investment – a residence – with less cost, you have created thousands of private equity dollars from the day you finish your house. The financial leverage that you will enjoy in the real estate market is significant. The built environment's asset value is approximately AUD 550 trillion. By contrast, the world's gross domestic product is AUD 110 trillion.

The Industry is Not Going Away

Construction provides two of four necessities to human life, i.e., shelter, food, clothing and water. Contrastingly, many manufacturers have left this country. However, construction cannot be exported. It must be "in situ" or occurring where it produces the product. Unlike service centres, computer programming or engineering, other countries provide it from afar and then send it back to Australia. Construction and its sister, demolition, are captive to the site.

Merit-Based

The construction industry rewards hard work. There is no substitute. We are all dissatisfied with the work ethic today. When we find it, we reward it? For example, a person comes to you (male/female/other) who does not speak English well but has promised to work hard. You give them a chance and one year later, you are glad you did. They kept their promise. Now, what will you do? Ignore them? Cut their pay? Of course not! You will increase their wages and give them more responsibility. Construction contractors reward merit.

Having earned the technical understanding and crew following, that same person might start their own business in a few years. It is almost expected. It is normal and rational to attempt it after working on the site, interacting with clients and managing labour.

Additionally, the industry is also merit-based. Question: what the best advertising in the construction business is – a completed project on time and on budget. It speaks volumes of a contractor's savvy and diligence. There is a minority of sound contractors. Word of mouth will travel fast. Excellent contractors have more opportunities for work than their lesser competitors.

Small is Big

Construction rewards the small construction firm. That is, they make a higher percentage of profit. This is a variable cost business. In other words, you do not have to have a "critical mass" to be profitable. Net profit statistics consistently show that smaller contractors make a higher profit percentage before tax than their larger competitors. But, of course, smaller projects have larger margins. Construction is an industry where the big do not eat the small, but the fast eat the slow. To be fair, the statistics show that small to medium enterprises (SMEs) are a significant part of the construction industry, employing approximately 65% of the workforce and contributing 50% to the industry value-added. By contrast, large enterprises employ 15% of the workforce and contribute 25% to the industry value added (ABS 2022). Counts of Australian Businesses, including Entries and Exits, July 2017 – June 2021 | Australian Bureau of Statistics (abs.gov.au).

Tangible

Our industry erects monuments. Our work is visible to everyone. We can see it for decades after completing it. Construction people show friends, relatives and potential clients these projects each day. Unlike other industries, we know what we accomplish. Likewise, there is little room for puffery, i.e., "smoke and mirrors".

Highly Paid

The construction industry for non-supervisory production work pays the third-highest wage of all industries. Overall, employee compensation in the Australian construction industry is third also.

Best Earning Years Later in Life

Statistically, this is difficult to prove. However, anecdotally, we have seen enough financial statements and compensation plans to believe this is a fact firmly. The prime earning years appear to reside in the 1950s and early 1960s. In our experience, the older contractors do have competitive secrets and keep them to themselves. It is to their profitable advantage. As a trend, a contractor's earnings grow over the years. We have seen increases as either a percentage or gross dollars or both. Their staff's salary and bonuses tend to increase as well. Construction becomes more profitable with experience.

Why? Because the business is about people and processes. A senior contractor has had plenty of experience with both.

People: he knows more people just by the years he has been in business. He also knows which ones to coddle, chastise, ignore or put an arm around to motivate.

Processes: experienced contractors know what does not work and what does. They have certainly tried a lot of good ideas over the years. They simply know what works and what does not. Additionally, they do not make the same costly mistakes that a younger, less experienced constructor makes. Hence, they have a competitive advantage.

In contrast, the best earning years of most other industries are between 35 and 50 (Professional Sports and Fashion Modelling excluded). The reason for this early, high compensation is the energy, willingness to travel, take risks and make extraordinary things happen. Subsequently, bonuses are increased. In contrast, profitable construction is based on consistent and correct processes, somewhat like the manufacturing business. The more consistently a person does the correct things, the better the outcome.

Age discrimination is an issue. Relatives and friends who work in various industries may say that they are underemployed. They all have advanced degrees and are over

50 years old but are not asked about higher positions in their firms. It is sad to see and even worse to experience.

To re-emphasise, construction contracting consists of two components – people and processes. An older executive does have a deep understanding and experience of handling people and building projects (They have the scar tissue to prove it!). My conclusion is a constructor's best earning years tend to be in their 1950s and 1960s.

No Consolidations or Mergers

The construction industry is an owner-operator business. Efforts to consolidate have shown the small business' power; you cannot beat an owner who is risking their wealth every day. Large firms are at a disadvantage. As of this printing, the highest market share of any one company is 2%. Again, small is big, and the fast eat the slow. This translates into thousands of family owned businesses that will continue to operate if another generation is willing and able to take over.

Local

Where can a construction company be started? Anywhere people circulate or live. Construction expertise is needed in all parts of Australia. A firm does not require a port facility or wide-open spaces to operate. You can start one where you live and that is family-friendly. This is an essential consideration for all working professionals who want to stay closer to their family and friends.

All Educational Levels are Welcome

Construction people who have worked onsite for many years have the equivalent of a college degree, of course, informally. However, most construction knowledge is learned while working, not studying. Technically, you know how to install quality work with your two hands. University students studying construction do not go to class wearing their tool belts. We passionately believe that someone who has successfully owned and operated a construction business for 20 years has an advanced degree. However, college graduates will earn more and have better career prospects.

No Large Capital Investment Needed

As is the legend, some successful construction firms have been started with little start-up capital. In this business, beginning cash is not a significant obstacle. This is a cash flow and variable cost business. To start a contracting firm, a person does not have to float a public share offering or have a rich uncle. What they need is an understanding of economics and construction craft skills.

Shortages of People Wanting to be in this Profession

Industry economists agree that we still have a shortage of people employed in construction. As we have learned in our lifetimes, the Supply/Demand dynamic is powerful. A lack of anything drives the price up. Surveys have shown that people in construction do not recommend our industry to their children. A significant association recently queried its members with the following question: "Would you recommend this industry to your son or daughter?" The answer came back as 72% said "no".

Competent construction professionals' earnings have outstripped general wage increases in other industries. That is job security as well as wealth building. Ask a computer programmer or an airline pilot about people's oversupply and the effect on wages and opportunity. You may remember the pilot overabundance driving down salaries in the 1990s. It is a danger to any professional; however, there is no danger to construction people in the foreseeable future.

In summary, what other industry has all these attributes? The answer is none. Although, the perception persists that our industry does not have much to offer young people. It is not true and a myth.

As an industry strategy, we should simply communicate these facts to the interested parties. Let us challenge each of us to promote our industry. Let the truth be told! And it is a remarkable story. This graduation season may be the best time to have a chat with your young people.

Career Advice to Young Constructors

We have been asked for career advice from young construction professionals on several occasions. So, here is the start of the conversation that others should continue. We have collected and listed 20 observations for young construction professionals to consider. We are sure there are many more if you ask any executive in a construction firm. But, again, here is a starting point.

What are the foundational items never to forget regardless of the career stage for a constructor? The young constructor could ask the same question – what skills and activities are critical to focus? Here are some hard-won lessons which not to forget.

Limit Working for Your Family

Your siblings, parents and other relatives have a mixture of motivations and biases regarding a young constructor family member. Many anecdotes can be shared here.

> Since you are a family member, you are typically less demanding, including pay. As a result, you are part of the economic model for their profits and not the recipient of the free market assessment of your value.

> They will be forgiven more often for egregious comments and behaviour. Families forgive, but do not make it part of your professional journey.

> Their memory of you is unfair, including your naïve child years.

> They will be too harsh or too forgiving of you. As a result, you will not grow as quickly as a constructor due to a lack of challenge or the stifling of your aspirations.

Every Constructor has a "Golden Lesson"

As a famous sports philosopher once said, "*you can observe a lot by watching*". Less speaking and quiet curiosity will illuminate efficient techniques and easy processes all constructors have – one or many.

Answer Your Harshest Critic

Those who have a demanding eye are often leverage points to eliminate our poor professional and personal habits. Many times, they are more honest about you than you are about yourself. However, thick skin is needed.

Develop a Few Non-Negotiable Ethics and Habits Then Improve Them

Your few approaches to the technical and business aspects of construction, including your careful routine that is safe, productive and has led to your success, should continually improve. This is how you consistently delight the boss and project stakeholders. These few habits should be sacred, thus, non-negotiable. In a rushed situation, you will not forget them.

Be Quiet for the First Month of Your New Employment

Perhaps the best advice before graduation is "*listen more than you speak, especially in the first month of the job*". What was meant was, you will not know anything for a while. Indeed, the politics and the company culture were alive before you got there. So you need to observe and not comment until you know. Hearing things from "*the new guy*" may result in resentment and damage your ability to work well with others.

We have seen over-enthusiasm hurt young graduates in their professional development. Much of this emanates from professional respect others may not have for you. You must earn it first.

Do the Basics Well

Be on time, take notes, follow up, among other practices. When you are new, you are a rookie. We all are rookies when we do something for the first time, at any age. However, doing the basics can only show that you are focused and that you can be trusted with more challenging tasks in the future.

Be Curious and Capture Knowledge

Your career is a long upstream swim in a winding river. You cannot see the end, but you can see the next turn in the stream. Nobody has a perfect journey. It takes a consistently intense effort to accomplish something substantial.

Be Cost-Conscious

This is a cost-side business, so this focus is essential. People who own construction firms usually have had some difficult financial times before the situation improved.

So, you are spending their money. Show respect for that and be a good steward. By the way, being frugal is not "cheap".

Be Supportive of Your Boss

Everyone has faults and anybody can be negative about any person. However, the person of good character is positive or silent, never negative about their boss. Other people you may work for will see this good attitude and will not forget it. Therefore, their plans may include you.

Think and Write Before Speaking

Clarify your thoughts. Know what you need to say and what others may say in response. Construction people challenge. Thinking aloud or "on your feet" is not a good habit.

Ask for Help Only After Making a Valiant Effort(s) to Solve an Issue

No one will disrespect you for trying extremely hard first and then seeking assistance. However, do not be afraid to ask questions or ask for help if you are in too deep. In contrast, do not ask for assistance first when assigned a task. As you know, your employer expects you to work independently. They hope you learned to work independently previously in school.

Take Slights With Grace

As the new person, you will be low in the informal pecking order and you may experience a hazing ritual that comes with being young or new in our complicated business. People may have a little fun at your expense. It is just part of the slightly abnormal humour we all share in construction. Older supervisors tend to do it to all young people. It is normal. Take with grace.

Be Modest

Someone is always better than you are. Also, some processes are not as elegant as yours might be, but they work. So be respectful of anything that works. As a side note, giving someone a deserved compliment does not take away anything from your professional standing.

Salt it Away

Save money from every paycheck. Even if it is only a tiny amount, you will be surprised about your accumulated amount when you become a senior professional. See the rule of 72-it rewards those who invest early and often.

Do Not Worry About Being Seen as Successful

Visible success is a myth. In construction, success is defined differently. From labourer to journeymen to operator to project manager to executive vice president, success is the same: you earn intelligent people's respect. They know how hard and unfair our industry is. Doing the things that have solid reasons behind them shows thoughtfulness. Persistently pursuing a solution shows character. These are non-visible but respected things.

Do Not Display Your "Certified Smart" Symbols

Take down your diplomas. Your education should be evident to others in your actions. What you know is more important than some paper that says you learned it, a.k.a. "certified smart". Considering that many do not have at least a bachelor's degree, you may be perceived as having contempt for others without formal education. Taversky and Kahneman assert that education is knowing what to do when you do not know a new or unfamiliar situation (Lewis 2017).

Seek Perfect Information

You may not have the developed skills or experience that older professionals may have; however, asking questions, researching and knowing more about any issue or project will place you in a "go-to" position. You will be sought because you know what the facts are. The construction industry has dozens of moving parts and a significant amount of uncertainty. Knowing more than most will give you greater value than those who guess or do not know the facts regardless of your age.

Work Smarter and Steadier

We have talked to many, many "20 something" construction professionals. They are brilliant and they feel they work extremely hard. Fair enough, but do not compare work ethic notes with a poverty born project manager or an immigrant site supervisor; you will be embarrassed. To our point, strive to work smarter and steadier. Look no further than the job site. The best journeymen and equipment operators work steadily. Our business is so complex; you must be steady in your effort. It allows you to catch the details and keeps you from rushing, which produces oversights and mistakes. Again, observe the best craftspeople and equipment operators for proof.

Your Body of Work Matters

After a couple of decades, your body of work will be well established. If it is solid, your career options will be many. If it is weak or inconsistent, your options will be fewer. A body of work operates on an elementary principle: your efforts, conscientiousness, skills and other like characteristics will "come out in the wash". You cannot fake a great professional body of work. This business will test you and find out who you are. It will make you humble or make you quit. So, build your skills and habits carefully and thoroughly. Do not take shortcuts and the lazy way out. People in this industry know all the excuses and they can recognise a fraud when they see one. The kind of professional you will deliver few or many options when you are our age.

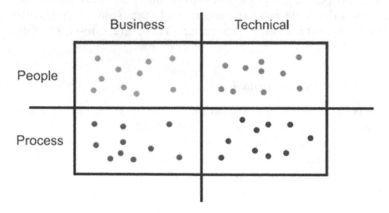

Figure 1.1 The Construction Contracting Knowledge Challenge.

The young Australian constructor's challenge is to master diverse subjects. See Figure 1.1. Many tertiary institutions teach many technical construction subjects and a lesser amount of business and career-oriented issues. This is a good balance, but it must know many functional areas for those who want to lead their companies and the industry. Figure 1.1 is a simple ("low resolution") categorisation of those areas.

Construction learning is a critical component to the young Australian Constructor. The new entries have been told the importance of leadership, management, and technology practices. Retirees exit with their out-of-date ideas, but recent graduates bring up-to-date thinking. They have "fresh eyes" and add to these areas. Overall, they have a positive effect on any construction firm.

We believe that improving the industry readiness of the new entrants is a meaningful discussion. This section suggests a limited number of focuses, as indicated by our industry interactions and classroom experience. Next, we discuss our reasoning and conclusions for recommending four general learning focuses and already credible learning, teaching and assessment processes commonly known as the Formative Approach.

This section is our culminating experience and combined thinking from two environments – 1) The Construction Contracting Industry and 2) The Built Environment Classroom – undergraduate and graduate. Our combined teaching experience covers more than two decades, but we do not pretend our knowledge is exhaustive. Others know more. We desire to add to their thoughts. We see construction learning as a journey that is never static nor complete. We intend to prompt more thinking and discussion about this vital industry subject. We will use undergraduate education for applying our assertions.

Our work with the industry has given us a helpful understanding of the contractor's desires for newly hired graduates' skills. We have talked with dozens of graduates whose titles include Assistant PM, Site Engineer, Financial Controller, Project Coordinators and Safety Managers. Regardless of their final educational degree – there appears to be an opportunity to improve this transformative process and focus.

Contractors will hire the majority of any construction program's graduates. This discussion is part of the healthy evolution of increasing the quality of those graduates. We are neither calling for new requirements nor an expanded curriculum, only a consolidation of instructional direction to what construction contractors appear to want. Our list of skills for the young Australian Constructor to capture starts with four *1) Vocabulary, 2) Conceptual frameworks, 3) Iterative problem solving and 4) Writing.*

The first focus listed is "Vocabulary", which starts the process. Students are immersed in construction labels and definitions. As many know, this helps instructors and students (and eventual employers) understand each other's questions and answers. Precise language is a value among most construction executives. The last focus listed, "Writing", finishes the process. A person with an emerging writing skill will grow their thinking and speaking performance. Looking past undergraduate education, it is no secret that writing is a core part of graduate school and for a good reason.

The Four Fundamentals to First Master in Our Industry

This section describes four focuses that young Australian constructors should adopt to progress professionally in the industry.

Vocabulary

Vocabulary is the set of proper labels and definitions of concepts. Words and actions are all we must interact with others; an excellent industry-ready vocabulary is critical for the new graduate's credibility and effectiveness. Precise vocabulary lends gravitas to the speaker. It gives them a perceived depth of understanding. Someone with a functioning industry vocabulary can communicate with fewer misunderstandings in less time. Everyone wins.

This vocabulary focus – the use of the right word or phrase – comes only from understanding the word or phrase's accurate meaning. To learn this is a remedial exercise – less exciting than technology improvements or risk strategies, but no less important. Words are a part of language, and without language we cannot think or communicate efficiently. Using precise vocabulary is less risky. Good words form the basis of accurate conceptual frameworks, which leads to our next focus.

Conceptual Frameworks

Having a "mental model" helps the professional construction filter and interpret what they see or hear. That same template can help them explain their idea. The first conceptual framework we suggest is that success in construction is always predicated on:

Safe behaviours
Quality installations
Cost adherence
Schedule timeliness

These have obvious value. Providing these four outcomes to Construction Service Buyers makes for better business relationships. Relating every issue to the possible effects on Safety-Quality-Cost-Schedule is a great "best practice".

We have shared graphic representations of these frameworks for young constructors to understand, such as Risk–Reward, Supply–Demand, Overhead Cost Application Target Marketing. We purposely inserted these graphics to quickly communicate essential considerations when deciding. As everyone has heard at least once, "a picture is worth 1,000 words". We count more than 40 different construction-contracting concepts that can be graphically displayed and labelled. It is important to state that we believe most construction professionals are visually attuned.

We have used this book and its graphics as the basis of the assessment of university students. As part of the final assessment, the instructor gives a list of concepts and asks students to sketch each, labelling all parts and describing their meaning and application in construction contracting. As you might guess, it is the rare student who scores a 95% or above. It is so few that the instructor can name each of those students. Affirming our belief in this graphical technique, several previous students have communicated that this approach has helped their careers. Many times, they use graphics to explain their thoughts better. From this, construction graduates might be well prepared to sketch their directions graphically to site craft, operator and management more often, leaving less ambiguity about their instructions. It is often directed at subordinates, sometimes to peers, but it has helped with superiors on a few occasions.

One graduate, Vibha, impressed her employer with her crisp understanding, including accompanying graphical explanations of 1) Margin versus Markup, 2) Project Return on Investment, 3) Construction Risk versus Reward 4) Strategic Planning, among other critical concepts. Indeed, middle managers received a refresher on these valuable ideas at the same time.

We have had several young graduates participate in a strategic planning session(s) with senior staff. Why? They knew the fundamentals of the construction business and were able to describe them clearly, sometimes drawing the graphics for others to understand. Their up-to-date knowledge helped keep "guardrails" on otherwise wandering discussions, along with their energy improved the meeting. In all, they assisted leaders in keeping the conversation relevant and moving forward.

Iterative Problem Solving

"The first answer is the wrong answer", one of our mentors says. Those who hurry to conclusions as if they are on a television game show are often wrong. Even if correct, they usually have not thought their answer through and it will not stand the test of time. Details matter in construction contracting.

Careful and often slower thinking facilitates better answers, decisions and qualitative judgment. Any construction challenge usually possesses a complex set of dynamics. Only when we question and re-question do we give ourselves time to collect and analyse more information carefully. As the last step, interjecting subjective considerations helps, so our conclusion(s) are holistically firm. Using case studies and challenging problems help show the value of iterative thinking for the instructor.

We use several classroom examples to demonstrate this. One is of a racetrack and a car's speed to show the need to think and compute first before answering. Another is a Return-on-Investment calculation challenge. Still, another is a quiz about the cost of a bat and a ball, together and separately. One of our favourites is a bridge-building exercise – a bidding and building competition – against other teams. We have used our books' cases and other's construction-centric ones from places such as Harvard

Business Review for case studies. Our instruction to students includes that famous saying, "work the problem".

The emergence of intercollegiate competitions in areas such as Design-Build and Heavy Civil has much the same effect. Each school's team is given a set of plans, specifications and scenarios to determine an approach. In a sequestered setting, they are left to think through the challenge with only each other. From our understanding, iterations of thought are in the dozens. This is realistic to problem-solving in construction. The judging panels are industry professionals who deliver the final iteration.

Iterative thinking is often slow but the end product will stand the test of time. It is a habit to instil and nurture. The resulting thoughtfulness and detail orientation are often high.

Writing

Writing leads to thinking and speaking. Iterative writing with feedback from a knowledgeable teacher clarifies thinking. If writing is improved then, we have seen improvement in the other two. However, well-expressed thinking is not as evident in the construction industry as contractors wish. This means that graduates who have clarity and organisation in their writing are more valuable than other applicants.

One compelling reason for focusing on writing, liability risk is high for construction contractors. Contemporaneous, clear and complete writing protects contractors. Quality writing is only possible if the previous three Focuses are ingrained. That is, vocabulary identifies, conceptual frameworks organise and iteration produces complete thinking. This sequence of learning has helped our students.

Writers must research to generate impactful expression. Whether it is delay history or re-mobilisation costs on a project, these facts must be discovered and vetted for truth before they are committed to paper. Lectures and questions start this research sensitivity on the student's first submittals in the classroom.

Organised expression earns creditability for the writer (and speaker). Their words, concepts and expressions deliver (or not) information to others that convinces, persuades, instructs or projects. Again, we observe that the quality of thinking and speaking follow.

Vigorous writing is centred on content, not the formatting or rules. However, errors are always a concern. To keep errors low in number, an editor – (or now, software) – can help. Content is many things, such as a fact, argument, insight or explanation. When the construction professional who is trained in writing is immersed in the attributes of a complex situation, clarity emerges and a reasoned path forward can be determined.

Everyone can learn to write competently. In our experience, this is true. We have used the example of writing a loved one's eulogy. Any person who has lost someone dear can write a compelling message about the departed. The focus is high and the deceased's essence is clear to the writer. Many times, there is no holding back of content. Due to the personal nature of such as task, the eulogy will be rewritten multiple times until it states what is intended

As an everyday example, if a person is passionate about Virtual Design and Construction (VDC) – mastering the four focuses is a pathway we have seen work. At the end of their training, writing about VDC's challenges or new developments clarifies their thinking about such things. To wit, they know what they think and why they believe it. In this writing exercise, they have mastered the vocabulary and conceptual frameworks. If their superior has time, they may make the young person re-write it, answering additional questions (placed in the paper's margins). These four skills are valuable assets. Writing helps clarity of thought begets insights. Insights may be breakthroughs that others have never. These four focuses transformation of thinking is core to what we think starts a construction career well.

Managing Labour is the New Opportunity

What is the most profitable construction contracting type? Let us ask it more directly: Which has larger profitability and more flexibility in recessionary and boom times? Our answer: labour-intensive contractors. Many young people think the builder is the only valuable career choice to make. However, we know that subcontractors are the ones that have more value to the industry.

The labour component of any construction project represents the most significant opportunity to improve safety, increase speed, lower cost and eliminate defects. Therefore, the line item on any job cost or profit/loss statement determines meeting, beating or failing any project's goals.

Since fewer qualified craftspeople are available than in the past, managing the existing pool more efficiently makes sense. Moreover, managing trade labour is a scarce skill and thus valuable.

We ask a fair question to everyone according to banking data: "Why is the average net profit percentage before tax (NPBT) of labour-intensive contractors double that of subcontract-intensive contractors?"

The main contractor's margins are limited. There are many reasons for this, including hyper-competition and the use of creditable pricing services. It is easier to ascertain the price of work. This happens regularly and makes clients cost-aware, thus profit limited. According to a list of publicly held construction firms posted by the Australian Stock Exchange, main contractors and construction managers have single-digit NPBT. Some suffered negative profit pre-COVID fiscal year ending 2019.

This is not so for a labour-intensive contractor. They may lose money the first year in business, but long term, as they grow a strong labour-management expertise and craft skill, they affect the cost side of the equation. Thus, their net profit percentage can reach double digits. Additionally, do not forget, this kind of contractor can generate service and maintenance turnover in slow economic times. So, long term, we like the second strategy. It makes for a more profitable business and an excellent retirement program.

Trade contractors have a complicated algorithm in their labour-intensive environment to arrive at a final price; few outsiders can put it together. Part of this algorithm resides in the mind of the construction service buyer. They desire a trouble-free installation during the construction and operation phases of any building or infrastructure project. Great craft skill gives a subcontractor more upward leeway flexibly, setting the price in the user's mind. Since it is somewhat rare, this translates into a more significant profit opportunity.

Also, as part of labour management skills, the risk of managing labour is rewarded. Risk in construction should be rewarded. Over the long term, if mastered, you will double your reward over a subcontract-centred business.

The business is about people and processes. Construction processes are taught very well, but what about construction people skills? The good news is that any weakness, if improved, has the most significant impact.

Being the type of manager who attracts and leads craftspeople and operators is the goal. Construction people talk to each other. Site professionals are no different. If a manager is perceived to be, fair and a good leader, others will seek to work for him, pay notwithstanding. Having a solid understanding of beneficial management behaviour takes time. Beginning this learning makes sense due to the many years it may take to become superior.

In general business, the subject of leadership has been well documented and discussed. This country and its stellar citizens have shared tremendously helpful insights about making things happen through others. Cannot we do the same with the construction craft and labour management?

The construction industry is the largest private employer in the country, and we have dozens of practitioners and professors in each city who could teach this skill. If the industry is fortunate, they will write textbooks about the subject as well.

Industry advisory boards are active with universities to ensure students' academic requirements are balanced with the industry's practicality needs. We assume the topic of construction labour management has come up in meetings; however, we have seen it in very few University curriculums. Why?

There are three times as many subcontractors' contractors as main contractors in Australia. The construction world is significantly affected by a construction firm's ability to build custom work in a unique place with equipment and people. A portion of construction always involves labour. Project owners are not immune. They must be labour sensitive if they want projects consistently built on a budget, on schedule with safety and quality (and without much conflict).

From others' analyses and ours, an average speciality contractor that generates a 10% increase in labour productivity doubles their net profit before tax. It is an even more excellent leverage point for equipment-intensive ("yellow iron") contractors as efficient operators affect equipment utilisation and cost. In addition, great productivity or labour savings transfers readily to project budget compliance and schedule adherence.

We do not know all universities' struggles to make this a valuable university unit; however, looking at standards and asking long-term questions is fair. In a broad sense, it is necessary if we are to grow as an industry.

Some programs will never be premier academic institutions as graded against others. This is the reality, just as any organisation or individual would be wise to be realistic about its/his ability and talent. We all have greatness in some things we do. However, it is also true that some people and organisations do many things well. Therefore, it is no shame and surprise that not everyone is in the "top ten".

Regardless of a tertiary institution's ranking against others, if they want to attract the right kind of attention from construction firms that will hire most of their graduates, the goal would be to make graduates "industry-ready". This is a vital strategic direction. Our industry loves and rewards practical skills.

It is noteworthy that all education levels are acceptable in our industry. Again, practical skills appear to be a common theme. According to ABS, there are approximately 38,500 construction firms. Additionally, of citizens as of 2022 aged 25 years and older, 28% have a bachelor's degree. Looking at the construction industry and applying this percentage, 72% of construction firm owners do not have a four-year university degree. If we can extrapolate, more than 275,000 construction firms are managed by someone who does not have a bachelors, masters or doctorate.

Most cost overruns occur in the labour arena. Craft skill and the efficient management of it differentiate contractors. Building and infrastructure owners want the work completed on time and budget, with safety and quality. Delivering this is a team effort, but site labour is where approximately 30% of construction costs reside (more if the equipment is considered operated by labour). It is the great "wild card" in meeting project goals.

We offer a direction. Here are our thoughts:

> Construction labour is different from the general labour pool in many ways. A significant percentage work from project to project, many times not in a continuous fashion between them. The hire, separation and rehire rate of over 50% of the total employment. Also, qualified tradespersons are rare and they have many opportunities to work for others, including overseas. They have the critical skill and must be served as well as managed. Many are not looking for the next rung on the corporate ladder but are certainly open to better pay and working conditions. They work to provide for their loved ones and themselves. Again, a different approach should be considered than a general business one.

> Analysing different generations working in the construction industry would make graduates more sensitive to managing their work styles.

> A DISC Assessment should be used in any course, teaching people skills. DISC is a creditable framework as a starting point in understanding people. It is a common, tenured and well-understood method. As an industry contribution, we have researched the relationship between DISC and the best craft people.

> Some cultural or ethnic information should be shared. Australian history is any guide; we will always have a significant percentage of immigrant labour.

> We suggest part of the course touch on the four basic types of managers. It is interesting to note that in construction, we should not be "coaches" to everyone. It would cause many problems, including lower productivity and higher risk. However, our research suggests that each manager needs to have the skills to be all four types for our diverse labour population.

> Any class might include the process of forecasting labour needs and availability for a construction firm's projects. This has a high value in the industry. It increases utilization and efficiency while keeping promises to clients. In addition, there is some anecdotal research that asserts it lessens stress and burnout to site staff.

> Students should understand the tremendous opportunities – financial, schedule, safety and quality – in learning how to manage construction trades and labour better. Therefore, case studies, statistics and quantitative analysis would be a significant part of this course.

We offer these as starting points for a potential syllabus. Indeed, any serious curriculum has many more. Labour is the means of production in our industry. We suggest construction labour is the elephant in the room and we need to talk about it.

Tertiary construction programs have made many contributions to the industry. They have partnered well with contractors and other industry professionals. Our clients and we are asking them to make one more contribution.

Managing construction labour is a rare skill. Therefore, learning associated with this critical piece of construction's overall process is sought. Managing people is not straightforward and parts might not be definable. However, we believe starting at an earlier point in a professional's career serves everyone's goals.

The Most Profitable Contractor Types

If you review available data about the most profitable types of construction firms, you might be surprised that it is not builders or main contractors. This is because they have the most turnover of any built environment organisation, including designers and suppliers. However, turnover does not equal net profit percentage.

As we stated, trade contractors generate the highest profit percentage, but which ones? There are many structural and market reasons for some to be more profitable than others. Here are the top three and why:

1   Civil Contractors. These organisations have three primary reasons for their high margins.

    a   Settled construction law guides disputes – if a project is stopped or delayed due to conditions or owner error, precedent directs the steps to resolution due to the many decades of cases arising from Australia's infrastructure construction. It is important to note that the government will likely be the project owner, follow precedent, and have conflict resolution systems for a speedy de-escalation.

    b   Construction equipment is often more expensive than the labour cost to operate it. This means firms may run multiple shifts or work overtime to complete work while saving thousands of dollars in job costs.

    c   Early completion bonuses are many times part of the contract agreement. Civil contractors will work extended hours – see above – and complete work in calendar days ahead of schedule.

2   Flooring Contractors

    a   Small contract bid profit percentages can be easily raised by one point or more without affecting the price's competitiveness. For example, calculate ten more percentage points on $1,800 versus $18 million. It is easier to justify the former increase versus the latter.

    b   Customers want to select the colours and textures themselves within a budget. Since there are many different types, qualities and colours of flooring which are very difficult to inventory in the local distributor warehouse. Many of these items have to be special ordered, so other contractors do not enjoy some profit opportunities in these areas.

    c   Typically, labour is contracted, so there are few areas of potential cost overrun.

d   Flooring manufacturers support their contractors more than other manufacturers by agreeing to inventory plans and quick ship programs.

3   Glass and Glazing Contractors

a   Glass and Glazing manufacturers have limited and protected contractor-
operating areas. For the exclusive representation and installation of the
manufacturer's product, competition is limited. Most products installed
accessible among all contractors such as concrete, timber, steel or
wallboard-not so with glass and glazing systems

b   Installation is contracted at times and thus risk of loss is limited.

Additionally, some may say roofing contractors are very profitable. That is true in our
analysis, but it depends on the season and the economy. They would rank in the
top 10. Some higher profit margin opportunities are due to re-roofing work, especially
in a slow economy. The other primary reason is storm work. When a region is de-
vastated, there are not as many roofing contractors as are needed; thus, price and
profit increase.

Success in Construction

We have studied success in construction for many years, reading and applying Welch,
Giuliani, Stockdale, Collins, Robbins, Covey and others. Each has written about it
extensively, but we feel there is more to add.

Commonalities, as well as differences, exist among the best professionals in every
industry. We know and have worked with hundreds of construction people. Our clients and their employees are a small sample considering that employment is somewhere above seven million. However, our data set is from several countries and dozens
of market sectors.

First, let us separate personal and professional success.

Personal Success

Personal success is between your ears. If you think you are successful, you are. Of
course, if you think you are not, you are correct. In human history, people from all
walks of life have made this determination about themselves. Some of them took
action and hurt themselves – slowly or quickly. The same is true to today. Alcohol,
drug abuse, food obsession and self-inflicted gunshot wounds are activities of the
negative individual. We all know someone who has many blessings but does not value
them. The reverse is true. Some people who live below the poverty line see the world
and their existence as an opportunity, full of possibilities and overall, a divine gift.

Changing a person's perception from negative to positive is almost impossible work.
The person must want to change; rarely can you make them see things differently. If
someone thinks they are successful, it is true. A visit to any developing country
confirms this. Not a great insight, but it must be said.

Does anyone besides us believe they are "personally successful"? Sure, most of us
do. Millions of construction professionals have dinner with their loved ones every
night, go to sporting events, music recitals and generally participate in the lives of
people they love. Many consider this success.

As a guide to our discussion, let us start by reviewing how one group of people – new immigrants – defines success. As a side note, all native-born Australians would be wise to review the two points below:

* **Provide for your family.** Family is primary and most new immigrants view providing food, clothing and shelter for their loved ones, whether here or outside Australia, as their responsibility.
* **Keep your family together, close to and communicate with regularly.** The desire is to have your family with you. Sometimes it is not possible. This has nothing to do with cars, clothes or memberships. It has everything to do with your ancestors, successors and your role in life.

Judging whether someone is personally successful is an offensive exercise. It considers our personal bias. Each of us shades facts due to our experience and hindsight. The same goes for those who judge you and your actions. To make it simpler, let us all agree to worry about ourselves. To be frank, it only matters what you do. A friend once told me, "Worry about yourself; it is a full-time job".

Professional Success

As we studied professional success, we found that many personally successful construction people are excellent at their jobs. No surprise. Statistically, many perform their function better than 75% of others in the same position. They are the "go-to" people.

How does one accomplish this? Anyone can by focusing on continuous improvement. The most common approaches are by studying, observing and only plain working hard. Savvy individuals "compete with themselves". They do not worry about how others might be doing. They know what they do matters. Twenty years later, if you challenged yourself against your personal best, it will primarily determine your professional effectiveness. Furthermore, we also see this as a dramatic example of the rule of cause and effect.

Here are five foundations of professional success in construction:

> **Successful construction professionals worry about controllable events** and spend minimal time on uncontrollable ones. As our world is more complicated than 20 years, this is especially important. We can worry about how things are changing, but we can continue to search for items that help us improve our technical and human skills. Competing against your personal best is one of the controllable. In 20 years of this, it is predictable that you will have improved in every facet of your life.

> **Successful construction professionals are independent thinkers.** They take time to quietly think through a problem or issue. They resist making a snap decision. Indeed, they seek counsel from bright and trusted friends. However, they believe in detail about the reasons they should or should not do something. Emotion is absent. It includes considering unintended consequences that may bite them later.

> **Successful construction professionals know there is no "100%" solution or perfect answer.** So, they manage the negative of any direction they take. Without

exception, high-performing people do the best they can and then "turn the page," focusing on tomorrow's challenges.

Successful construction professionals believe the Stockdale Paradox. Former Vietnam prisoner of war Jim Stockdale practiced two principles while surviving their ordeal 1) He was faithful to survive through their talent and focus. 2) He was brutally honest with himself about their situation. Contrast this to others who were overly optimistic to themselves about "being home by Christmas" and some even died after being broken-hearted several times.

In Jim Collins' book, "Good to Great", successful companies expressed faith that they could figure out any situation. However, they all were frank internally about what they did well and did not do well regardless of whose feelings it might hurt. Effective managers mainly consider decisions addressing "what is weak", not "whom will it hurt". Subsequently, each "great" firm worked the hours needed to solve problems or reach goals and moved personnel that was best for the team and maybe not the individuals involved.

Successful professionals know people have a more certain future than companies. Companies come and go for reasons small and large, but there is always a need to hire superior professionals. They are rare and are paid better. There is market competition in pay packages to attract these people. It happens each day. Shelter, processing facilities and infrastructure must be built. At over 90% employment in Australia, there is no economic recession for capable professionals.

There are several more but, these are our five foundations for being professionally accomplished.

We have all heard the term "best of class" contractors. It is a consulting term that has been thrown around like the word "superstar". However, what it means in the mainstream press and what it should mean are two different things.

"Best of class" contractors are judged by financial ratios achieved in their business. That is, they are financially successful. There is no consideration for technical knowledge. The "best of class" construction firm does well fiscally but not necessarily in craftsmanship. There is some, but not complete correlation as to technical construction ability.

In our experience, many financially successful contractors are great business partners to their subcontractors by leading, coordinating and managing construction projects with them in mind. There are dozens of other reasons that we do not have space here to discuss. With that, the best subcontractors focus on working for these superior main contractors, builders and construction managers.

Overall, construction professionals do not mind the demanding client; however, they appreciate the fair client. "Hard but fair" describes the best owners, funders and users to work with. Projects get built very well and outsiders wonder how the construction team does it consistently. This is one of many reasons.

So, would you agree we should start using "financially successful" and stop using "best of class"? Financial success is about money management and not about building work. In these days of political correctness, the former title accurately describes the fiscally savvy construction firm. However, it does not guarantee it can construct with higher quality or faster than a craft-able or personally successful one.

Professional success is not a fuzzy concept. We conclude it means you are better at what you do than three-quarters of others who perform the same job, whether a project manager, site supervisor, foreperson or labourer; being excellent has many rewards, including personal satisfaction.

We believe success is not a number or a title. Numbers can be fudged and there is always someone who posts a better one. We know that it is not about riches. Titles betray a person's contribution and promote the idea of position as necessary. All these denote competition and thus emphasise winning over others at any cost. Of course, some unlucky people fail to reach their goals not based on controllable activities such as effort, discipline, foresight or planning.

We need a description that can be used in all circumstances. Here is our one-sentence definition:

Professional Success is the Respect of Intelligent People

These knowledgeable and experienced people know how unfair and demanding the construction industry is. In other words, intelligent people know your challenges and appreciate your approach. You have a reason for everything you do. Others might say, "you have figured it out". These intelligent people are not opinionated or ego-centric folks who do not work in the construction industry and know our business. However, they are the smartest people in the country (just ask them).

We demonstrate the above as we respect many people in the industry for their perspective, reason, intellect and hard work regardless of title or wealth. In our opinion, "*the respect of intelligent people*" gets to the core of professional success.

Taking Off the Toolbelt: The Transition to Manager

The site is where a construction contractor makes a profit. The project manager and site supervisor's quality is linked to the level of gross profit a contractor enjoys. Each site supervisor is a resource coordinator, taking the job's demands and matching them with limited resources. At the end of a project, the site supervisor who thinks more as a manager will win. To win more often, this supervisor must psychologically take off their tool belt and assume the role of business owner. The business is the construction project. The average construction project in Australia is about the same as the turn-over of a small business. The site supervisor is responsible for it as a small business owner is. Here are some profitable strategies in ensuring the site supervisor is an effective leader and manager:

People are Fragile

Deep down inside, your workers want to be taken seriously. It is the number one human need in a professional setting – respect. Take them seriously. As a starting point:

a   Take time with each member of your crew to teach and understand them.
b   Never use their nickname. People will never be offended by using their given name or a shortened version of it.

c   Paychecks should be handed out with a "thank you for your work this week".
d   Always assume that people are trying to do an excellent job until proven differently.
e   Make sure your people have what they need or, at least, fight to get it.

Because Things are True Does Not Mean They Need to be Said

If it is accurate and positive, indeed say it. However, keep the negative comments out of your conversations. It keeps the atmosphere professional and forward-thinking. You will be tempted to state an obvious shortcoming of someone. Do not unless you are on your way to fire them. Then, it is still a hazard.

Well Organised Beats Smart Every Time

Success in construction is based on a good, thoughtful process executed on every job every time. Daily organisational skill makes anyone look smart. Planning weekly, training people in small increments, knowing the contract documents and keeping material constantly in the work area are typical of above-average forepersons. Conversely, intelligent, disorganised people in construction are a plurality and not well compensated nor respected.

The Process is the Problem, Not the People

The speed of construction, number of people, differing site conditions and conflicting goals cause problems on most projects. Your team comes to work wanting to succeed. They would like a raise, be considered for promotion and go to the next "big" job. Success comes from having a thoughtful, proactive process.

You are Buying People's Energy, Ideas and Enthusiasm, Not Their Time

Do the things that keep people in a positive state of mind. Being appreciative of your journeymen will encourage them to follow you. Having a sense of humour never hurt anyone's outlook. Giving a well-deserved compliment will not damage your leadership. However, being a pretender will come out in the wash.

People Want to be Included in the Plan

Thirty years ago, people just wanted to receive their instruction and go to work. Today's typical worker wants to know what to do and why he should do it. So, include them in the things that are appropriate to share.

Talk About Bad Situations Only Once

Yes, strongly voice your disappointment behind closed doors. Make it a point of showing your displeasure. Use the opportunity to raise the performance bar. However, once the door reopens and you walk out, that is the end of the verbal spanking.

Money Does Not Buy Happiness

We know this, but your people will constantly talk about their paycheck. Why? Because it is easily measured and hard to argue. What do they sincerely want? To be appreciated, to go to a good job and to have a bright future. In surveys, compensation ranks in the middle of all variables concerning job satisfaction. You guessed it, being included in the company's plans along with appreciation and a bright future rank ahead of pay.

Do Not Think Out Loud

In other words, do not tell people what you are considering or might do. This confuses the crew and starts rumours. Instead, you state a direction you have chosen and will not change your mind when you speak. Do not voice your thinking and you will keep your crew's rumours and distractions to a minimum.

Do Not Forget to Tie Any Raise to Improved Performance or Skill

For open shop situations, raises should be mentioned as a concept. State this in the months preceding the raised date. The smart foreperson discusses ways to maximise the raise by acquiring skills or showing improvements, such as competent person training, working independently or installing so much work in a day. When raise time comes then, nothing else needs to be said. You either give them a minimal or maximum raise.

It is the Little Things That People Notice

Remembering their concerns and interests tells them they work for a good person. Take a good coach, for example. He knows several things about each player and mentions them. Even when he cuts someone, they do not take issue with it. They know it is not personal. He cares about them as people. Take another look at all these guidelines. Sincerity and caring win with people. Yes, even if you cut someone.

People Work for People, Not for Companies

Most great contractors are great people. If you are someone whom they believe in, they will stay. The loyalty is there. As an aside, it is clear that most great contractors are "mud on the boots" types. The crews onsite have great respect for someone who has done their jobs. A friend said it better, "you cannot run a construction company by email".

Compete with Yourself, Not Others

Stellar site supervisors do not worry about other projects or their peers. Instead, they compete against themselves and focus on that. For example, if it takes them an hour to do something, they try to finish it in 55 minutes. Frankly, they do not concern themselves with others.

Get Ugly Early

When you see a problem, identify it and communicate it. No one ever lost respect for having the courage to speak of a problem ahead. Once everyone calms down, their minds will start thinking of a plan "B". Early detection always gives people more time to research and consider options. Having to make difficult decisions in an hour is a recipe for subsequent problems down the road.

Values Equal Actions Equal Culture

When a supervisor believes in a value, it is expressed in actions. People can say anything, but what they do speaks their heart. For example, if you value hard work, you do not care if the hard worker is your least popular worker; you mention it. If you appreciate courtesies, they are not taken for granted. If someone fetches a tool or a cup of coffee, you thank them for it. If you value organised habits, you notice them and point them out to others.

An actual value(s) builds a culture in your team. Indeed, when raise or promotion time comes, you speak your values through actions. Conversely, if you have limited values, your people will do all manner of things. If a site supervisor is weak or "laid back", their team will not focus. Thus, he will not produce. Pity the site supervisor in an industry that only rewards safety, quality, schedule and cost. He may have a difficult time growing in their responsibilities.

Break up Cliques

Allowing cliques on the job site does little to enhance teamwork and learning. As we move people around to work with others, we do a service for ourselves and our people. A new partner adds perspective to performing tasks in different ways. Also, the practice gives a person an understanding of different personalities. In summary, you are giving a gift. They will gain an insightful feel for people and personalities. It is inexpensive management training for your next site supervisor.

Can we ever know too much about managing people? Can we ever have too many human resource skills in our toolbox? We think not. Most of us will never completely understand the mysteries of people or how to manage them.

One part of managing people better is the ability to teach them. In this industry of labour and management shortages, it is doubly important that we can take raw talent and mould it into something better. That is the only option left. Many alternatives, including raising pay unnecessarily, have been short term solutions but overall failures.

A Coaching Parable

Excellent teachers and leaders in the world use parables. Some religious texts contain hundreds of them. Yet, using a story to teach is still unconventional in construction. Nevertheless, any measure does not dispute the parable's power. Parables are an integral part of Aboriginal, Micronesian, Indian, Australian and many other societies' cultures-they are used mainly in directing and teaching young people.

We have used two effective strategies to make our staffs think:

1 Asking questions; and
2 Telling parables.

Management stories are a quick way to direct all people in a company. To define, parables are stories that have a point or what some people call a moral. So why is the parable a great tool to direct and teach your people? Simply put, because:

They are indirect and non-threatening to an individual.

They make people think about a specific issue; and

They can be repeated over time and will become a short-hand management code.

It may take a couple of "tellings" to your people before they comprehend the reason for the parable. As the leader, you do not have to explain. After repeating the story three times, most normal people will conclude an underlying message and realise the point.

Many stories such as those presented here are available and applicable to the construction and management themes. You just must find them or create them.

The "F" Words Needed to Manage in the Construction Industry

The use of the "F" word is historical in war and construction contracting. Here are the "F" words to successfully manage your construction firm.

Friendliness – Relationships and people grow in a friendly atmosphere. All people do seek social atmospheres and turn away from unfriendly ones. So let your place of business have the air of positiveness and not the tension of negativity.

Your managers and supervisors will undoubtedly look forward to coming to work. They are halfway to wanting to do an excellent job for you. See the other four "F" 's below.

Frankness – It pays to be frank. However, it is uncomfortable for everyone in the first half-hour, the healthy discussion of what happened and why makes your people better performers. Of course, be friendly in the process.

Fairness – don't play favourites. Always seek "what is the right thing to do" versus "who is right" this keeps all your employees and other people you work with treated equally. They will always return the favour to you and treat you fairly.

Faithfulness– Always assume that your people and business partners have the intent and capability to do a good job. Do this until you are proven otherwise, and then be frank. After that discussion, forget the past and turn the page into a blank one.

Of course, if they do not have the intent and capability to do a good job, look below.

Firmness – Have minimum performance standards, including behaviour and quality, that apply to all people. Use the other four "F" s to communicate this. In the long term, you will polish your reputation and performance, although you will have some late nights in the short term.

Parable: The Championship Season

This first parable is a general one that most construction people will understand quickly. A coach and their team won the championship. Their season record was 18–0 and afterwards, the town organised a large victory celebration, including dinner, to honour the coach, the team and their accomplishment.

After dinner, the coach approached the lectern to speak. He could not help but notice all the time and effort exerted to organise this event. As he reached the podium, he asked, "Would you have gone to all this time and trouble if we had gone 0–18 instead of 18–0?" A voice in the room shot back, "No and we would have missed you, coach!"

Meaning: It is expected to have talent and translate it into wins (or profits). Unfocused and but gifted managers need to be goaded into thinking about being better. Being talented and losing games (or money) is of no use to the boosters (or the company).

Another application of this story is that contractors constantly fight complacency and the entitlement mentality of some of their people. This story addresses this kind of frustrating issue. You should not start by threatening action but by delivering this message. This lets more competent employees correct their behaviour while saving face.

2 The Business of Construction Contracting

Managing a construction firm well takes more than just creating a first business plan. However, not performing this step ignores the unpredictable winds that will undoubtedly blow each year. If a young constructor does not plan, they react and any decision made quickly has some adverse effects. Sometimes, an unintended change of direction is due to the company owner's distraction. These winds and distractions drive a firm to an unintended place without a formal plan, often to the business' peril. Therefore, revisit and revise your plan each year to create contingency plans; focusing on details is crucial. It is your insurance against falling prey to unexpected circumstances.

There are building blocks of scaffolding a good plan into action. Some first-time firm owners may not know the industry dynamics, while others do not have a trusted framework to plan. Their understanding is anecdotal and not a mental model. It is tough to make sense of this industry from the limited number of experiences. To be influential leaders and managers, most people need a way of thinking that survives their multiple mental tests. Behavioural Economists Tversky and Kahneman noted that an expert's thinking model is better than the expert's judgment (Lewis 2017).

However, it is critical to note that management fads are. They are often fuelled by employees who do not want to become victims of the next fad. They know it is better to be an internal advocate than a dinosaur. The consultants who formulate and sell management fads such as best practices are the initiators with a powerful monetary incentive. Transcendent definitions of best practices will forever remain elusive; they shift due to the people's focus advocating them. There are many diverse groups in construction. It is essential to state that best practices have all the characteristics of a management fad (Green 2011). Remove the word "best", and companies have a balanced perspective with less arguing over ideas. It is sometimes easier to improve the world than prove an idea's value (Lewis 2017).

The timely completion of construction management practices varies from company to company, and the lower the value of a method, the lower the productivity (Gurmu 2016).

In our observations, construction company employees do well when defining and stating their firms' best practices. In the company, its operations and strategy can create prioritisation from most value to least value. What works for one firm may not work for another. The implementation order has to follow the company's prioritisation.

Applying a mental model is an efficient method to engage in a complex process. The construction contracting industry is complicated. An example of a usable

DOI: 10.1201/9781003290643-2

mental model is to view the construction company on a continuum. An inter-connected chain of events. How you start out is large reflects how you will finish. Lax work discipline with ill-defined practices and profits will be unpredictable. The company will have less value when the owner exits. Due to the loose operating nature, they may involuntarily exit.

On the continuum, the interconnection between the quality of work and clients is positive. If you perform subpar work, a quiet word of mouth will circulate amongst construction service buyers. However, good work done promptly will also be circulated and most project owners are open to entertaining another proposal from a quality firm. Reversing poor word of mouth is difficult and time-consuming. See the ship metaphor below.

The continuum framework can be likened to a large ship sailing in the open ocean. There are many places to stop and be distracted by project owners who want competition and not necessary you as a construction service provider. However, there is an ultimately valuable and meaningful destination for the leader. They should understand what that destination is and proceed with focus. If the owner wishes to manage a less clear and more relaxed way (this is not a criticism), the recommended approach might be to keep the business small and focus on those types of projects. Higher margins with less risk can make for an excellent work/life balance.

The building blocks and mental frameworks in this chapter are meant to aid you in interpreting the construction industry effectively. Many times, we will use graphics to illustrate clearly and quickly. This is intentional. Most construction people are very visually attuned.

A Basic Profile of a Profitable Construction Contractor

Many contractors entered the construction business without much capital, trained people or management expertise. Yet, today, some of these same contractors are leaders in building infrastructure, residences, office buildings, schools and infra-structure. They are the "go-to" people to construct with quality and safety at a fair price completed timely. Personally, these same contractors have attained financial security and a professional reputation without a superior.

NASA has quantified many of the qualities needed to be an astronaut. Candidates must have the "right stuff" to be accepted in the space program.

What is the "right stuff" to be a successful contractor? We can think of all the successful contractors, and it seems they have little in common. Some perform com-mercial work while others focus on residential and still more work in the industrial field. One is union; another is not. Some prosper in metropolitan areas and some in rural locations. Some have autocratic styles of management, while others use a par-ticipatory manner.

To be clear, there is no profile and we would be wrong if we suggested one. However, successful contractors are better than average at one thing:

Managing Resource

Construction is a cost-side business. Containing cost is the only place of focus to deliver margin. As you well know, most Contractors cannot receive a premium of even

1% over their peers. They have to price work at the market while delivering the same quality as those same peers.

This means that there is a little cushion for inefficiency or mistakes. Added to that is the demand for speed. It is no secret; clients want the finished product faster than just a few years ago.

The contractor has to rely on (mainly) the same labour pool. Also, he is limited to using the same local material suppliers that everyone else uses. So, there is no significant advantage there.

The contractor has to be excellent at managing resources to meet all these demands. It is the only variable. Therefore, keeping costs contained, primarily labour expenses, is critical.

What is the critical activity to being a superior resource manager?

Gathering Accurate Information

Whether it is labour productivity, project leads or return on working capital, timely and correct information is crucial to making the right decisions. Without it, one contractor is ordinary. So, seeking and confirming information is an essential skill. Some of this data comes from the market and some of it from inside their company.

A contractor will always have a better result and more confidence knowing he made an informed decision.

Computers help in gathering and analysing information. However, this does not mean computers are the only tool for being informed. They are essential, but they are as good as the data received and the analysis model used. In some cases, paper and pencil calculations are far better.

Take, for instance, labour productivity. A contractor is far ahead if their field manager can quickly compare labour efficiency on a daily timesheet rather than waiting on a multi-page computer report from the office.

Tendering models are based on accurate information. Practitioners know that using them consistently leads to added profit margin in the long term. Some people use electronic spreadsheets, while others use a paper method. All beat the competition.

We could cite many other examples. Our book has many of them. However, our industry observation is that excellent contractors are also superior resource managers who make decisions using accurate information, no matter what tools they use.

Manufacturing has its value for society as an economic engine, but so does construction. However, as many have commented, there is room for improvement for the built environment in many ways. Research on its complex nature continues and should help it deliver even more value.

A Starting Business Model

The most general depiction of the construction business dynamic is below. These are the processes that all construction businesses must perform if they are profitably pursuing a continuing existence.

The components of the construction contracting business are shown in Figure 2.1. These four functions every construction contractor must master. Each through the life of the firm should become faster and more accurate. The apparent areas that slow the cycle down are the gaps between each part. In some cases, information may

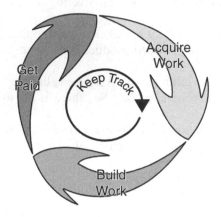

Figure 2.1 The Business Cycle: Safety, Quality, Cost and Speed are a Competitive Edge.

be transferred inaccurate, thus causing rework and even reversing the direction of the process for a moment.

The graphic in Figure 2.1 represents a part of a value of the mental model we have mentioned in this book's forward. Mental models help integrate the chaos, change and new information. It is our internal model of how the world works (Senge 1990).

Information should flow well in your firm if the cycle increases its speed. Outsiders, including clients and potential buyers, will objectively review your capability will look for 1) people who are cross-trained; 2) financial and schedule information flow to be a top priority; 3) disbursement of information to be faster than data collection and 4) use of mainly electronic means to distribute. Having these done well and promptly will produce a faster business cycle than your competition.

Figure 2.1 reflects a new business model in Construction Contracting. It is from the culmination of changes over many years. Some of these changes are fair and some unfair. However, there are some new conditions under which contractors work. The business has changed over the last three decades. As an example, all projects are now fast track. Contracting is more about the company than the craft (clients have forced us to change). Legal professionals are increasingly litigious and there are more of them. Designers are under the same cost and time pressures as contractors, hence less complete drawings and specifications. The work ethic is not as expected. This paragraph could be must longer. To a normal person, the construction industry is not a rational place. There are hundreds of market segments. No two construction organisations are the same. Each market segment is defined by geography. Individual regions where work is put in place have issues to contend with such as local codes, inspectors and political problems, which cause frustration when an outside construction company enters it. Specialised trade skills define most contractors. That skill is what contractors trade with clients to receive income. Each contracting firm is unique.

The owner usually has grown the business from zero turnover to its present-day volume. They, along the way, have made thousands of decisions of how the work will be built, how it will be estimated/priced and how it will be tracked. This series of decisions give the construction firm a unique way of conducting its business. As a result, the construction industry has multiple moving parts.

There are thousands of steps to build one project. They can be done in any sequence. The outcome will be a disaster but, steps are not ordered. It is only the project team that determines the order. If they have an exemplary sequence of thoughtful steps, they will win more often (speed, cost, quality and safety). The contractor has hundreds of steps to execute on an individual project and in the business for the contractor's part. Again, they can be done in any order; however, the above-average contractor insists on mapping the major steps in a good sequence and executing those steps. Discipline has to be part of the new business model in Construction Contracting.

Principals of the New Model

1 The construction contracting business is comprised of people, processes and a hierarchical structure. All things built involve these three components. Therefore, constructors who think about these three major leverage points of the contracting business will enjoy consistently profitable projects.
2 The business of construction contracting is now a tennis match. Each side (contractor and client) must continue to put the ball in play. The game must be stopped if one side stops returning shots (such as not responding to RFI's, or variation notifications). There is no room for continuing to build and then settling up at the end of the project. The risk is too significant. The expectation of professionalism should be high. Both parties must respond to each other.
3 Dedicate energy towards the challenges that can be controlled or influenced. Do not invest time into things that your actions cannot affect, such as the weather.
4 Volume for the sake of volume is no longer an option. Size doesn't matter. A smaller contractor will have a higher profit percentage than a larger one. Therefore, a contractor must find a comfortable volume for them.
5 The risk/reward curve is inverse to the general business curve. This means that projects will a higher risk have lower rewards. Not surprisingly, lower-risk projects have a higher ROI. Therefore, understanding risk-reducing practices allows projects to be more profitable over time.
6 Finding the "sweet spot" on each contractor's risk/reward curve is a quantitative exercise. There are a small number of quantitative drivers that place each contractor in their correct place. No longer should we gain this place exclusively by experience. This is inefficient and dangerous to the business.
7 The process of building a project is complicated. There are dozens of moving parts (decisions and steps). Added to that, everyone learned from someone else. The process is the problem. There is no "GAAP" for constructing methods. Define the process among the project parties and the job builds faster, safer, with higher quality and less cost.
8 There is little that is normal in construction. Contractors are doing many things differently. Clients are more demanding in requirements than a decade ago. As a result, contractors have some leeway in the way they approach problems as owners want solutions. Young people can contribute to this area due to their new thinking
9 Construction companies should use thoughtful business and technical processes. That should be one defined process. There is no prohibition to doing so. However,

taking time to map out, document and monitor will be an investment in the future that pays off.

10 Best practices (BP) are the best way of starting. Of course, any best practices list is not 100% appropriate for any construction company. However, it creates a conversation about what works and what does not, forcing a company to determine what is best for them.

11 Any company benefits from using a good process. On any single project, a contractor may have problems. That is to be expected over a 30-year career. However, it is no reason for a contractor to stop using best practices. BP does increase the odds of success.

12 Simplicity means speed. Simpler processes are quicker and easier to teach than elegant ones, thus more efficient in working with other people.

13 Each contractor must find a niche(s) to fend off competitors. The industry has enormous competition. According to the Australian Business of Statistics, there is approximately a 6% increase in competitors annually. These new companies tend to work for less and often say "yes" to your clients. A solid strategic action is to build a high wall around your current clients. Make sure they know all that you do for them. Raise the bar of what they should expect from any new entrant. Some of this is just communicating all that you do for your clients. Some of this is adding a further benefit. Of course, the former is less expensive.

The new business model is based on maximising profits and building a valuable business in which to sell. Increasing value is directly related to the tangibles that contractors deliver to the clients:

Safety
Cost
Speed
Quality

Gaining a reputation for delivering this is the ultimate value position that places the contractor in the pole position with the client. All that is needed is a new type of thinking. The changes of the last half-century indeed suggest it.

There is no one solution in construction. Try as we may, it will always take several approaches working in concert to make a project come in on time and cost with safety while keeping a construction contractor profitable. An older constructor has learned many hard-won lessons on keeping their business and her projects running smoothly. they will be the first to tell you that there is no singular approach. "It depends" is not a cop-out but a truthful answer.

This is the answer of the constructor who has earned the practical MBA. Through hard knocks, lessons learned, scar tissue and generally overcoming several crises in their career, he has learned "situational management". When the first set of facts is discovered, it helps us think of a solution but not determine our final answer. The second set (the other shoe dropping) tells us what to do. However, let us share several observations and a few conclusions about this trend with you. This trend is one for the ages, not just for the decade.

First, Let Share the General Ideas and Then More Specific Thoughts

If someone is selling software, TQM, recruiting or other services, they may see their offering as the answer to construction's problems. We see these valuable services as part of the answer but, these solutions must be balanced with core needs. Following one of these as a total focus is to ignore the complexities of running a construction firm.

We have an industry that is sick of a thousand cuts. Many little things compose most of our construction ills. We have all searched the world to answer construction's challenges and some of us are still searching. Suppose you find that one great thing that answers all of our challenges, you will earn a boatload of dollars selling it. You should be well compensated; it is the rarest of rare in our business.

As multiple approaches are the answer, we have categorised them into three main groups. Any profitable construction business is based on three keystones. All else is subordinate:

Processes,

Compliance to Those Processes,

Risk/Reward Curve Position.

Processes are those things that work in making the cost and turnover of a project distant from each other. Thoughtful processes drive costs down mainly through improved human resource productivity and asset utilisation. Good processes keep corporate expenses lower than the competition for a construction company.

We are in a cost side business, and the greater the cost distance from the revenue, the more secure the business. Said another way, if these two business numbers are close together, the company may have one destructive event and cost can quickly become more than turnover (a loss).

Some companies we know have spent time creating, observing and documenting these "best practices", Subsequently, most have seen significant improvement.

As our background is in construction, we know the power of an exemplary process. One that is simple can be as powerful as an elegant one. We believe simplicity means speed and a straightforward practice is easy to monitor—an essential point for any manager or executive.

We will share with you several dozen practices in this book. You will be rewarded by reading them. They may not be new to you but, these practices will constantly prompt you to think of a better way to run your firm. In another way, we may confirm your current practice's value. Telling you that your process is a standard approach among constructors encourages you to keep using it and not get distracted by the latest fad.

Do not underestimate the power of quiet thinking as you try to solve for financial independence. There is always a better way. It is incredible to us that there are people who think working hard is a physical endeavour. We believe it is both a physical and mental endeavour. An owner or manager makes more money per hour by thinking (such as creating fast business processes and negotiating with others.) than their physical activity (such as business meetings or wearing their tool belts).

A firm's excellent and decent employees can help here. They have seen things that work for other companies. Either as an observer or while they worked for them. It does not matter. The process helps all people perform better. However, you cannot rely on good people to perform well if others sync with them. These days, all members of your team need to be dovetailing their actions with each other.

So, the process is critical. The process drives the other two parts of a trinity to profitability in our current environment. Without thoughtful approaches, compliance is moot and your risk/reward curve position is dangerous.

Some processes are tactical (daily) and some are strategic (long term). The daily ones can be documented in a business management manual and the long-term ones were written in a strategic plan.

Predicting higher or lower productivity compared to estimate is one of the essential steps to consider when building a model of job-site efficiency. Prediction appears to be based on the levels of planning and implementation. The adherence level of construction management practices is associated with low productivity.

Compliance with those processes means high predictability in behaviours. Just as a construction company is well regarded when it scrupulously keeps its promises to clients, so is the payoff when everyone inside a company follows through on their "promise" to follow the company processes—the faster and more accurate (discipline) the compliance, the lower the cost of business. The right employees affect this key variable positively.

Compliance is critical as the contracting market does not allow even modest premiums over competitors. It is a cost side business. That is where higher profits are made.

We have seen adherence as high as 90% in all processes from our work with firms in the construction business. These are highly profitable firms. You may think that this is a surprise that compliance is not over 90% or an "A". It should not be. Construction firms are the "tail on the dog" in many ways. They work for a funder and user (owner) and are subject to a designer's (Engineer or Architect) vision. Both parties tend to have their agenda and a highly productive contractor on their projects is not a top concern. As an example, insistence on a joint formal and thorough pre-job planning process is still rare.

The highest we have observed in a contractor's office is 91% compliance. We have asked some stellar contractors what their compliance is on a company-wide basis – on time and done right the first time – and the answer has come back in the 70% range. These are very thoughtful contractors whom we would say are the best at what they do.

Good and decent people are critical to any compliance improvement. These are employees who seek to improve past themselves. If their compliance is 80%, they strive for better. It is innate. To find these people makes your job of management much more manageable.

These people care about the company. In many cases, the company is you, the owner or the manager. If you are a good and decent person, you will attract and retain good and decent people. Leadership training can only take you so far. Character shows and if a good character is not evident, people can see it on the first day on the job.

Your Risk/Reward Curve Position is where the business of construction contracting comes together. Find the sweet spot on that curve and your company is at a place where there is a high reward for the risk you are taking.

The Risk/Reward curve is inverse to the general business curve. Lessen risk and you will make more money. It is odd but true. Increase risk and you will make less money on average. Long term, this is accurate. The effort of managing risk has a positive reward.

Some of which are unknown, i.e., who will next manage this country, what an employee may do on the job site regardless of your policy, an owner's financial situation and the like.

Lower risk construction situations (profit impacts) tend to have higher rewards (ROI). Companies that follow a thoughtful process do move upward in profitability because they lessen or eliminate these profit impacts. It is essential to note some of these risks; thus, your position on the curve is chosen by you. These choices are clients, markets, projects, locations, methods, products and many others. Saying no or "pass" at times is a good business decision. To be fair, most risk in construction is organic. Jokingly, some say constructing is Latin for "high effort and low reward".

These factors 1) Processes, 2) Adherence and 3) Risk/Reward Position drive profitability in construction contracting. What is comforting is that the business can be more quantitatively managed. We do not believe in a bloodless business model that discourages people from creative and entrepreneurial thinking. However, we feel strongly that this approach makes cost, schedule, quality and safety more predictable while driving them in the right direction. As a result, business relationships with clients flourish. No greater goals exist in the business of construction contracting.

Strategic Planning for Construction Contractors

We offer a tailored strategic planning model and process for construction contractors. Over the last several years, we have created one in our work with clients. Part of our effort is in reaction to what we observe; many construction companies use general business approaches such as Strength, Weakness, Opportunities and Threats (SWOT) or Political, Economic, Social, Technological, Legal and Environmental (PESTLE) Analysis. However, these are promising approaches; each analysis a contractor's current state and not a pathway to the desired state accompanied by the incremental steps needed to get there. Instead, GRAET plots a path and employs a method to reach that desired future state. We think it raises the efficacy of strategic planning efforts with construction organisations.

This section explains our framework and process to those construction firms wanting to utilise strategic planning more effectively. In our opinion, all strategic planning is more effective if an appropriate framework and realistic approach are used. These two inputs help guide the effort of multiple people, including people unfamiliar with strategic planning. Fundamentally, GRAET attempts to direct all efforts in one direction and becomes a shorthand to effectively communicate within the team during the many months needed to create and execute a strategic plan.

Below, we explain our solution. This is the product of much iterative work, study and long conversations since the 1990s. The model and process have two fundamental components critical to any strategic approach.

Transformation – Changing a company's reality to its proper "North Star". North Star is a term used by many to denote a company's core purpose. But, as you may know, it should not change significantly during the life of the company.

Journey – The slow march towards a valued operational state. We say slowly because each step needs to be done entirely while bringing employees' knowledge along.

Good strategic plans address significant internal and external components such as locations, project types, clients, markets and logistics. These components increase valuable outcomes such as reputation, profits or employee skillsets when improved or better aligned with duties involved.

Mintzberg et al. assert that there are ten different approaches regarding strategic planning. The different types include learning, planning and methodical (2005). The authors' multifaceted and nuanced approaches may confuse our discussion. However, it is little wonder that Strategic Planning often defaults to SWOT or PESTLE – for simplicity. More to the point, our proposed model reflects what we think is the germane approach to construction contracting from the universe of Strategic Thinking. We offer our GRAET model as an improvement to the current thinking. Its acronym lists the framework and guides the process.

G – Goal(s) of company owners can be many things. Here is the range of possibilities.

Time Required: *0 months* – should be evident now.

> To make a living so that one can provide for their self and loved ones. Indeed, paying oneself a paycheck as part of a business venture seems obvious. However, some owners have not extended their thinking onto other more global goals.

> To provide for our loved ones with a job. Keeping family close to a parent while working may be unspoken, but a real motivation for working without a passion for the construction business.

> To build monuments. Indeed, many constructors fell in love with the construction business with its physical nature. This may evidence itself by targeting types of projects or the market share of certain project types.

> To make a profit. For large firms, this might be turnover and % of net profit goal by region.

> To build value, then leverage that value for personal and professional options. The person who believes this may have any of the goals a-d. However, they have extended their thinking to the end of their career.

Indeed, there are many iterations, combinations and nuances concerning the strategic goals listed above.

R – The Reality of the company and industry

The time needed: *One to six months*.

As a general statement, an effective strategic plan deals with controllable and uncontrollable events. We can plan the controllable and try to influence the uncontrollable. Most time is spent on the former. In this step, we use SWOT as an analysis framework. An honest assessment of reality is what emerges.

This is SWOT's value; to help objectively guide thinking about a company's business. Certainly, assessing the quality and predictability of its current performance and projecting its sustainability into the future is critical.

This crucial section helps create genuine strategic insights. From all the facts gathered with the owner's goals in mind, connections can be made that competitors may never realise. Therefore, it is worth the work involved.

Reality is best assessed by seeking data and facts. Indeed, source data (not other's analysis) such as government statistics, banking data and association research can only be interpreted insightfully by you or your managers for your unique business situation. Some time must be spent thinking about what the data means and its impact on the future. Here are a few common areas to review.

Internally

Financial Statements and reports give us a ready-made look at the company. Labour, Material, Equipment and other direct costs are captured. Only after this is analysed compared to peer competitors nationally and regionally, that comparison can be confronting but healthy.

Compliance to Practices. This is another core reality check. Statistically, we can measure with some accuracy the company's compliance with its best practices.

Site – Office Conflict is something inherent in all construction firms. If there is a significant one, this reality must be addressed. To have it is to slow the business process, affecting the four core underpinnings of any construction firm: Safety, Quality, Schedule and Cost. To achieve excellent results in these areas takes the full collaboration of the site management and office staff.

The metric of Tendering Gross Profit versus Market Share over three years gives a clearer picture of the trajectory of the company's business.

Externally

Risk Management – Do we have an overreliance on a client or vendor? Are we performing a risk assessment at the beginning of each job?

Demographic Shifts – are more people moving into our markets? Do this and other shifts lend more to renovation/repair versus new construction.

In construction, marketing finds, selling converts, bids confirm project price, terms and conditions and operations delights and keep clients.

What is the ratio of competitors per million dollars of work versus other markets or geographic areas?

More politicisation of the construction process. Are more construction dollars being regulated by local, state or national government? What is the ratio of Government employees per million dollars of work?

Less qualified craft people are available. What is the trend? How might it affect our construction operation?

Increasingly complex supply chain. Locally versus internationally sourced products and equipment? Do we need to lock up a portion of the production of a particular manufacturer? Are pre-fabrication and material inventory beneficial approaches?

Decreasing productivity. What is this trend – flat or dramatically reducing? What is controllable?

Partnerships should not be done for the sake of partnering. Both parties should have a need and both should be able to fill each other's needs. With that in mind, partnerships must be structured to be fair and if fair, they will be repeated in the long term. If alliances are organised, they will go much smoother and have more value to each side than earlier ones. So, again, they will continue and benefit each party.

A – Align people, processes and company infrastructure to goal(s) and realities

The time needed: *Six to 18 months*.

> *People.* Examples include hiring, mentoring and culture. Hiring those with measurable grit as part of the criteria can change the productivity of first-year employees. Mentoring grows employees' professional capabilities, including flexible thinking (important as changing conditions will continue). A culture that is centred on planning thinking rather than schedule setting is desired. Planning leads to less stress, crises and conflict among construction firms they work with long term, companies and their employees' benefit.
>
> *Practices.* We observe that standardised documented methods continue to be a missed opportunity for most construction firms. This standardisation assists companies to become more efficient since compliance with procedures is easier to improve project results.
>
> *Company Infrastructure.* Choices include Branch Offices locations, Software and Hardware purposing and selections and Hierarchical structure. Benefits and drawbacks must be analysed. What if scenarios are helpful here.

E – Enhance the people, processes and company infrastructure

The time needed: *18 months to five years*.

The specific focus is to make each of these better, i.e., faster, more accurate, safer, less costly, higher quality, longer life cycle or more valuable to your client. Here are some examples:

> *People* – hiring more people who are measurably "gritty".
>
> *Processes* – breaking down all functions into practices for straightforward teaching, learning and monitoring
>
> *Company Infrastructure* – looking at hardware and software, determining its dependability and how well it fits your practices.

T – Transform the industry

The time needed: *Five to ten years*.

In this last step, the PESTLE framework helps prompt people to think about the opportunities in transforming the areas it represents. Below are assorted other areas that could be areas of industry change.

Our software and hardware become standard with most clients, suppliers and other critical third parties.

Minimum financial strength and other requirements make for a higher level of competence and other qualifications standards for all clients to require.

Higher quality of installation. Assessments that ensure that the quality is second only to safety by clients and regulatory agencies.

Improved safety standards – who can be against perfect safety, including wear injuries?

Certainly, dozens of possibilities exist. If one review all the areas that affect the quality of earnings in a construction firm, it may number over 100.

Time

Time is sometimes overlooked as a critical ingredient in strategic planning. Since the complete execution of each strategic action is vital, small steps towards a significant goal over several years are often the best choice. Once a step is completed in the correct order, there is no rework or going back to re-execute this step.

When in a Crisis, A Short-Term Focus is Rational

In this situation, careful operations are more valuable than long-term strategy. When planning to start a business, there can only be speculation about what will work well and not how the market will respond to your service offering, how your employees will behave or whether your clients will fill all their contractual obligations. To start, a business plan is enough. There are more critical factors to consider in the beginning. Time will tell and events will occur. This will prompt better thinking about beneficial adjustments.

Long term, time "thickens components" or allows for components to align well and correctly. In addition, people become more familiar with them, mastering them better than first-year users.

Lastly, time allows for insights to develop. As each of us knows, honest and hardworking contractors are typically more profitable than in their first year. The time allowed them to garner insights about the efficient operations and overall value of daily actions and strategic directions.

The Ultimate Strategic Goal

From our work, thinking and observations. It is clear to us that there is a best strategic goal. In a sentence, strive to appeal to the most sophisticated client(s) in their chosen markets. Why focus on a high-quality construction service buyer (CSB)? We suggest four compelling reasons:

They have money during a recession.

They buy based on value and not low prices.

They will only work with a limited number of contractors.

They influence other clients and third parties.

We have seen companies move forward with this type of thinking and enjoy the benefits of relationships with savvy customers. In some cases, a sophisticated CSB mentor an honest, hardworking contractor and leads them to think in ways that deliver excellent safety, quality, schedule and cost. But, of course, the CSB does this for their benefit.

Conclusion

Construction is a unique industry that exposes a contractor's strategic goals to many uncontrollable forces. Companies have their unique operating characteristics with which to manage these forces. This reality must be addressed if an effective strategic plan is to be created.

We have seen enough strategic planning performed by general business consultants to have a creditable opinion. Through the years, our work and reading have been extensive. From our perspective, using SWOT or PESTLE as a core tool for strategic thinking is inadequate. These are analysis frameworks and not transformative processes.

In our mind, using a stepped transformation method that includes a timing component for increasing difficult change – from alignment internally to transformations externally – is rational. GRAET is our answer.

Our framework does not suffer from that frustration by a process that is unclear or weakens over time. Each component is labelled and is significant. Middle managers, who must understand and carry out strategic actions, can quickly comprehend what the ten-year pathway is and why.

In all, our method may be a "disrupter" to orthodox thinking. We do not propose this model to make an unexpected change but a substantial improvement to the largest private industry in Australia.

Whatever the reason, the question remains: "How will contractors address predictable change – proactively with a plan or reactively with troubleshooting – as they attempt to achieve their company's goals?" We believe the company's answer is provided by its year-in and year-out results and resulting market value when the owner transitions to retirement.

12 Critical Areas in Strategic Planning in Construction Contracting

We have witnessed many strategic planning engagements, most of which have resulted in profitable journeys. From afar and through working with construction contractors, we have observed these efforts. We have invested more than 40 years each in the construction industry as we have listed a dozen significant points to consider for strategic planning in construction contracting firms.

To start this analysis, let us revisit the reason for strategic planning. The compelling business reason to put forth the effort is to direct the company to a planned and desired end. Contrast this to the alternative: being buffeted and bounced to an unexpected place where you continue to be at the mercy of clients, markets, governments, suppliers and other outside forces. As an analogy, you are a passenger in a truck and trust the driver to deliver to a beneficial and safe destination. In contrast, a constructor's attitude should be that life is too short and the construction industry too risky to let outsiders determine their business destiny. He wants to be the driver of that truck.

To be clear, your company should have some history before undertaking your strategic planning exercise. Having built projects, paid bills, met payrolls and overcome a crisis or two will give you some hard-won insights into how the industry treats you and your firm and what you should do to make it behave to your benefit. However, if you are a first-year firm, a business plan is enough.

Here are a dozen cautions as you consider endeavouring on a defined strategic path(s). It is important to note that these are controllable issues:

> Sometimes a consultant is necessary to facilitate this process. One of the values of a strategic planning consultant is that they offer an objective view of the company and its plan. Their value is that the consultant reigns in over-enthusiasm; other times, they bring a better approach used in the industry. However, there can be traps to using a consultant. Here are a few:
>
> * They should possess some construction-specific experience. Generalists will use overall business models' ineffective plans. Our construction contracting business is unique. such as SWOT and concepts to guide you to an inefficient and
> * It may not help if the consultant is a recent ex-contractor. They may be too close to and have too much experience in the construction industry to be objective.
> * If they work only with your market niche, your plan may be like others.
> * Indeed, any duplication takes away from your plan. You may want to ask what companies they have worked with across the country and when.
> * All in all, a consultant's helpful attitude goes a long way to assist any contractor's strategic planning. If they take care of large and small details, this will make the process one where your people can focus on the critical issues and not the grunt work.
>
> Make certain past efforts are considered in the process. A contractor's failures and successes show where a company should or should not concentrate its efforts. There is no shame in reviewing missteps.

In most things, if you quit, you fail. However, if you continue to work at it, you will improve and eventually achieve your goal. It is important to remember that strategic planning works. You just must find a way it works for you.

If you have had success, look at the reason(s). For example, it might have been because certain people were involved or were the facilitator. Whatever works is the key here; do not change whatever works for you.

> Give the potential directions time to air out in between meetings. Enthusiasm is excellent in implementing the plan but can lead to errors in the creation of the program.

In our industry, we have many capable and aggressive people. We are not timid as a group. Make sure ideas do not have the problem of "we can do that!" at the meeting and "How will we do it?" later.

Sometimes a cool hand is needed as ego might be part of the motivation for a direction. The second in command – a consultant or even a spouse – may be that required sobering influence.

> All strategic directions must be vetted against potential financial performance. Be conservative in your projections, as money is the only barrier between you and bankruptcy.

Let the financial equation be your ultimate critic. If you cannot financially make it happen, do not do it. Borrowing significant funds has bitten contractors in the past. Our margins are slim and the construction economy is the first to slow. Add to that the myriad of risk factors and subsequently, the bank debt may go unpaid, leading to significant adverse effects.

As a simple example, starting a new service offering can be a daunting financial question. A conservative and safe approach would be to have six months of expenses saved for this new venture. As you know, clients may not pay you, but you must pay for everything. So, taking a fiscally conservative approach to a new venture is rational. But, again, our profit percentages do not allow for grand experiments.

One of the primary decisions for a construction firm is determining what processes and business areas to keep static and what ones to innovate. Organisations cannot innovate all functions at once. Gateway computers, with its dual technology and retail sales strategies, is one example.

More importantly, a growth initiative makes innovation in any area harder. You work with new employees, clients and project types while collecting higher contract turnover and variations as you grow. To ignore this and concentrate on improvements is harmful. In contrast, a static company (same revenue, same employees, same clients and same project types) can focus on innovation. It is easier to improve safety, quality and productivity with consistent clients, employees and project types. We have witnessed increased profit margins as a result.

> Always consider what reactions your competitors will have. Competitive moves against you must be planned for as your peers will react against your preliminary direction. As a side note, the most fun you can have in strategic planning is watching your competitors react, squirm and generally blink at your moves.

Indeed, people will start investigating your job sites and clients if they feel you are onto something profitable. However, some constructors advise not to give out information freely. For example, they may keep their signs off job sites and buy material from out-of-town suppliers. They make those who want to know to earn the knowledge. In the meantime, the contractor may have bought himself several months before the herd of competition enters the market.

A client from the West said it in a phrase: "*I make money alone in the dark*". Another from the Midwest says their firm is "*just rocking along*". Both know that not drawing attention to themselves is a profitable approach.

An additional benefit of this extra time can work on the cost of business. As other companies enter the market, the price will become an issue. However, if you have reduced your cost in this interim period, your proposal/bids can be less but with the same margin percentage.

> Your peoples' skills must drive the plan. If they cannot promote and sell the services of a new business unit, do not ask them. Additionally, be careful in hiring a new manager to lead that effort. Statistically, the odds are that this new employee (just as with all new employees) will not work out.

Depending on the economy, the right person might not be available. It is more likely to hire someone who does not quite fit your culture, although they look good on paper.

After the first mis-hire, companies become very sharp at what type of person they are looking for. As a result, the subsequent hire is a good fit in all aspects. Hence, the new venture is well-led.

You know your people will try to follow your demands. However, do not ask them to do something, not in their character or training. One practical management concept states, "pick the trained employee and not the one who will just try". Long term, we need the consistent effort of someone who understands the challenge of whatever the new venture is. Not the employee who will just try hard.

> Acquiring another business can be very efficient in growing turnover and service offerings in construction contracting. Construction companies are consistently excellent in their technical skills but are mediocre in their business processes. Sometimes, a company like this is open to merging with your firm. In the end, everyone is happy as you take over the office headaches and your firm acquires a new higher-margin skill set.

People get tired of our business and want to work regular hours with less stress. Family concerns may be the driving motivation. Health might be another. No matter the cause, keep an ear out for these situations.

Additionally, as we look at the highest value in construction, it is the qualified craft hour. Therefore, acquiring a firm with proven site people may allow you to enter a market without being the low bidder.

In another way, by acquiring a company, you can upgrade your craftsmanship. There is no great insight that quality journeypersons, operators and site supervisors take years to develop. However, increased production capability can be accomplished in short order with a company buyout.

The process must be quantitatively driven-not all emotion. The careful strategic process allows the data to prompt discussion. From that point, we are in a sane and sober place. Thinking through the resulting issues and insights will enable companies to outthink and outmanoeuvre their market peers. Again, from a place of reality.

This quantitative piece works as a safety net. You will not make significant mistakes. Your competitor, driven by emotion or ego but not data, may commit significant errors. As a result, he will suffer setbacks and reversals. As an immediate result, your competitive edge becomes greater.

Acquire good data and trust it. There are several industry providers and the government is a reliable source as well. Even if you must purchase quantitative information, it can be justified by an averted loss later.

Historically, some have mistakenly seen a small part of the market as a profitable place, aka "a niche". Contractors are sometimes goaded by an outsider, such as a salesperson or trusted advisor. Sometimes these niches are there; most times, they are not. The data will show you.

> One ultimate strategic plan takes a company's capability, finds a niche and then drives towards it.

Construction is a service business and cannot be patented. Therefore, even if a niche is solid, it cannot be easily protected from a competitor's discovery and intrusion. To define a perfect niche is one that is profitable and defendable. Three examples are: 1) Clients who seek resilient contracting companies. Financial discipline crucial to getting a significant financial capacity is not prevalent in this industry. 2) Work that requires certification or licensing. It deters people from investing much time to become approved. 3) Work that requires specialised equipment. Those who own it and are trained on it have an advantage over new entrants, have to rent/finance it and be trained to use it.

If you do discover a niche that is somewhat hard to protect but profitable, one strategy is to get in quickly with as much volume as you dare risk. If you know, others are just months away from following you and thus affecting margins. This makes sense. Many new competitors make a long-term profitable presence hard to manage. Some noise-abatement contractors are sensitive to this.

> A strategic plan is not an arms race to see how many initiatives and how many measures an organisation can list. Some younger practitioners have this characteristic in their plan (i.e., an impressively long list covering several areas). Great energy went into it. Lazy people did not produce the strategic plan and it came from hard work. However, will it be implemented?

When checking back the next year, all objectives might not have been reached. Also, you might find people less enthusiastic about strategic planning and the upcoming company workshop. The reason: the ongoing business concern is what employees must keep a vigilant eye on, i.e., what they were hired for. Employees are hired for the work at hand, not the strategy. Too many action items mean they all might not be achieved and those done are "pencil whipped" or little focus on the practicality or detailed implementation.

Keep the strategic plan and its actions limited. Of course, the directions must still improve the organisation, but they can be implemented modestly and less overwhelming.

Jack Welch asserts that any idea, however worthwhile, not implemented, has no value. That is, a million-dollar idea multiplied by 0% implementation has zero value and you may have wasted an entire year. Taking that same logic, a $100,000 idea implemented fully keeps its value. Indeed, a more robust direction may be added to it next year and even greater value can be realised.

One interpretation of this thought is that strategic planning can be approached from an incremental and iterative perspective. Small directions, constantly adjusted and refined, may be added over the years. In this way, a ten-year strategic plan is not over the top; it is reality.

All construction firms have made their share of strategic mistakes. However, an incremental approach with adjustments and refinements makes for minor (inexpensive) corrections and when these firms have finally followed the right path, the value has been noticeable.

Construction firms we know that have practised this incremental and iterative approach are better for it. However, their staff may be slightly impatient for the next strategic planning session – i.e., their energy level is not worn down (this is a good thing).

Our business has an enormous amount of detail and complication. We believe it has more than most and it has lower margins than most. From our work anecdotally, we assert that this modest approach works very well. Again, the implementation is where value is realised.

> Big Hairy Ideas (BHI) are exciting but long-term in nature. To think of a great idea is fun (entrepreneurial), but to execute is less fun (process). These kinds of new directions can take between 18 months (as in computer software and new service offerings) and five years (as in people).

Since implementation manifests an idea's value, choose your implementers carefully. Some people are easily distracted or discouraged. Those who are process-oriented are above-average in focusing on steps to completion.

> Great strategic plans look ahead to potential changes and events in the market. We have all seen them in hindsight – energy prices, labour shortages, legislative action, technology evolution and the like. Part of your strategic plan is to anticipate potential events and to plan for that contingency. A critical exercise is a SWOT analysis – the strengths, weaknesses, opportunities and threats of your business and its operating environment. Opportunities and threats are what you are looking ahead to and planning for.

Having a plan of action in case these occurrences arise can be a windfall. Unfortunately, most firms do not have a strategic plan. They will be slow to act and work out the problems of answering these threats and opportunities. You and your program have already thought it through. Anticipating events and having a contingency action process can make a difference in acquiring significant clients, valuable employees or percentage points of profit.

As a strategic aside, managing the documentation of your knowledge is valuable to the growing organisation, increasing the value of the firm and a windfall the successive generations who may lead the company. The goal may be to furnish a complete starting point for the subsequent owners of the firm. The simplest way is to keep disciplined, archived systems of information. One advanced but somewhat costly practice is the use of online archival systems. In addition, we advocate the storage of detailed post-job review reports. These reports are executive summaries that tell a straightforward story in brief terms. As you may have years of these, major themes, both good and bad, are practically all that is needed.

We believe our GRAET Model is more attuned to construction contractors. It is logical in its flow. 1) Set Goals, 2) Grasp the business, 3) Align the business, 4) Enhance the business and 5) Transform the market in which the company operates.

In conclusion, strategic planning is a critical process for an established firm. If yours is a young construction company, it will be something that should be considered in the second year. However, a business plan will suffice for now.

Keeping the strategic planning process from collapsing due to missteps (there are dozens) is controllable. We have listed only twelve. Amusingly, not performing strategic planning will eliminate failure. However, you will be stuck with a company driven by markets, customers, suppliers and other outside forces rather than by you and your knowledgeable staff. Take the time to do it and you will find a leverage point to launch you to the best places (such as clients, markets and geography) in the industry.

Professionals should treat strategic planning as the holy grail of management. It is powerful for those companies that utilise it. Done well, it differentiates a company and drives it to an unoccupied place in the market. As we stated, some firms are not ready for it; however, all should consider it. We firmly believe strategic planning is critical for any experienced firm. Figure 2.2 represents the company's position on the risk/reward curve and its direction of travel.

The Natural Barriers to Construction Innovation

The construction contracting industry has not been an active innovator. Research literature has noted that this is a historical problem. The rate of innovation and its adoption is low in the construction industry, as Zhang and Rischmoeller (2020) noted. However, new tools and technologies are recently emerging for the construction industry. This current phenomenon in construction is best summed up by a Bell Labs scientist, John Pierce, in 1948 – the advent to the computer era, "After growing wildly, innovation has reached its infancy" (Isaacson, 2014). The Innovators | Book by Walter Isaacson | Official Publisher Page | Simon & Schuster (simonandschuster.com). There is a real need to improve construction outcomes since it dramatically affects the quality of life.

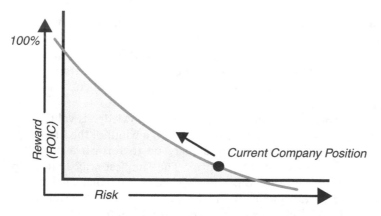

Figure 2.2 Organisational Risk/Reward Curve.

The industry has recognised this and moved to increase the quality of outcomes. The challenge is to overcome the many barriers that dampen it.

The construction industry is a privately held business sector based on profit and loss. It is not government based that makes space exploration such as NASA. As a contrast, Construction, just like a Silicon Valley enterprise, is focused on profitable strategies and systems, where NASA's success is found in the act of discovery in outer space, with cost being a secondary consideration. It is good to remember a timeless construction innovation joke, the beaver and rabbit view Gordon Dam in Tasmania. The beaver answered the rabbit's question with, "No, I did not build it, but it was based on an idea of mine".

There are unique challenges for construction firms to perform R&D. This section will outline the construction industry's key barriers and constraints in pursuing construction innovation. Many have noted that individual factors present in construction are not unique; however, the combination of factors is found in no other industry. In addition, there are significant underlying economic and market barriers that are structural and, therefore, difficult to change. Unfortunately, the economic realities of construction are primarily ignored by those who advocate management improving recipes (Green 2011). Table 2.1 refers to the Australian publicly held firms by selected industry: net profit before tax percentage.

Exceptionally low percent of profit for the risk pf this industry. The shelter, infrastructure and process facility sectors invest in research and development much less than other parts of the economy. These sectors invest less than 0.5% of the value of their sales in R&D, while the Australian national average is approximately 4% (Hassell et al., 2009). Summary of Federal Construction, Building, and Housing Related Research and Development in FY1999 (dtic.mil). However, our industry's reinvestment in R&D may be misleading. The net profit before tax of the average contractor is low compared to firms in R&D intensive industries. The large firms in the construction sector (namely the contractors, e.g., Simonds; Lendlease and Global Construction) net profit before tax is less than 10%, whereas technology companies are more than 10% and in some cases range between 20% and 30%.

Lumpy asset problem. The investment needed to research and develop a product or service is a "lumpy asset". This is a financial term defining a particular type of investment that can only be made in whole or not at all. A firm cannot "rent" or pay for using a lumpy asset on an incremental basis. The industry is not readily scalable. It is still a craft and location-dependent business. In practice, once the investment decision is made to pursue a specific research and development goal, it must make in total. So, additional and typically significant contract turnover must be generated to recoup costs and realise a profit from this kind of dollar investment. With the construction industry's economic uncertainty, making a significant research and development investment today appears to increase the firm's future risk.

Table 2.1 Australian Publicly Held Firms by Industry: net profit before tax percentage

Construction	Technology	Medical
Global Construction 8.1%	OFX 21.0%	Ansell 35.1%
Lendlease 6.4%	Technology One 21.2%	Sonic Healthcare 11.2%
Simonds Group 0.5%	Telstra 13.7%	Zenitas 13.6%

Source: Australian Stock Exchange–latest available data.

Extremely low market share for industry leaders. The largest construction companies in Australia do not dominate the industry's market share. For example, Lendlease has a market share of 2.0%, whereas ABN and Bechtel Australia have less than 1.0%, as noted in IBISWorld in 2018). Additionally, 98.8% of construction firms employ less than 19 people (ABS 2022). Counts of Australian Businesses, including Entries and Exits, July 2017 – June 2021 | Australian Bureau of Statistics (abs.gov.au). Market dominance has never occurred for any one firm. Firms such as Google, Telstra and Amazon have a majority market share in the areas they pursue. There are four major market groupings in construction, including 1) those who coordinate the work such as main contractors and construction managers 2) those that physically build the work such as subcontractors and specialist contractors,3) those that fabricate and systems and 4) those that supply materials. There are numerous supply chains with multiple markets within each tier and each level has various market structures.

Large number of competitors. Construction continues to be the industry with the highest number of businesses in Australia in 2018, accounting for 16.6% or 383,326 of all enterprises (ABS 2018).

The "intersectionality" of construction firms. Classifying construction business homogenously is problematic. Each business' is significantly affected by its characteristics and its operating environment. If categorising them by 1) project type 2) trade focus 3) region(s) operating in 4) client types 5) contract type(s) working under 6) publicly or privately held 7) amount and type of technology used 8) the number of employees 9) financial result 10) management structure. Since there are multiple answers for each of these answers, over 1 million combinations are possible. Due to the number of construction firms in Australia, there appear to be few similar construction firms. Any innovation may be having to general to improve the operations of many firms. Few buyers of a construction-specific innovation make it unaffordable for innovators to produce and buyers to purchase.

The industry is the first to slow in recessions historically. The growth of construction company turnover is difficult to retain due to economic cycles and the extraordinarily competitive and aggressive industries. There is little ongoing business from the previous year or customers who continuously trade in many firms because the volume of work depends on project number and size. In many situations, there is little cash and assets retained in the company.

IBISWorld (2018) identified Key Success Factors (KSF) for a construction business. The top three most significant include: "1) Ability to expand and curtail operations rapidly in line with market demand. 2) Operators must quickly alter labour force numbers to match short-term cycles in market demand. 3) The ability to hire experienced, productive workers, especially during periods of low labour availability, is crucial to success". These point to a short-term focus required by shareholders and executives of construction firms.

Construction firms tend to build several projects at one time. Due to this, project demand varies throughout the year. In addition, as mentioned above, construction is one of the first industries to slow down when the economy moves into a recession period. Therefore, profitability is uncertain because of these short-term cycles. As a result, the construction industry has a high failure rate when compared to other sectors.

The construction industry has the highest proportion of independent contractors of any business sector in Australia. The ABS asserts that the percentage is 37%, whereas the next highest – administrative and support services are slightly over 20%. A better-negotiated work scope and price is needed to increase asset productivity's input/ output ratio. This tactic seems to lessen the incentive to make employees more efficient and, thus, less need to create or adopt innovation.

The highest personal or top 3 corporate investment manifests itself in tremendous price pressure that keeps margins depressed.

Service is difficult to patent. Australian patents effectively protect unique tangible products for a legally prescribed 20-year period. However, this can be a protracted and challenging process that is a high risk to the patents' creator. Past research by London and Silva (2013) indicates the challenges for those in the construction industry to create and protect their patents. The Australian system affords few rights to the creator of patents and little protection with the onus of patent protection solely with the creator. As a result, actors involved in innovation who are novices can experience high risk. Coupled with this, patenting a process or protecting it from competitors' duplication is not easy. Employees move to other firms often, twice the rate as manufacturing, with the freedom to implement their knowledge in the next company. It is generally accepted that R&D investment and subsequent patents facilitate market dominance in many product-oriented industries such as medical devices, pharmaceuticals and general manufacturing.

High turnover of the workforce. Depending on the year, over 50% and approaching 100% over the total employment in construction. This means that the knowledge of innovation may travel with a departing employee(s).

Most **contractors are on a personal guarantee with suppliers and banks.** As a result, most of the owners of the approximate 385.000 construction firms must guarantee the re-payment with their personal assets to qualify for a credit line.

As an aside, we note that there appears to be some *regulatory malaise*. For example, the government is neither proactively updating building codes nor business rules that govern innovation in the construction industry. Unfortunately, we cannot offer any data to this point but, we can say that governments do not reach out to the industry and examine where the government could assist or grant minimal approval to encourage promising ideas to be developed.

This barrier combination is unique to construction; therefore, we try to adapt other industries' proven innovations for our industry. Many do not fit easily. The above constraints slow those that do. Regardless, once any solution is confirmed as valuable, it still must be tailored to a specific construction firm that implements it. University research outcomes are typically publicly available for anyone in the industry if the government funds the research; unless separate agreements are developed and approved between the institution and the company. If the research and innovation is a process and business solution, it may be quite challenging to customise the outcome and directly transfer it to implementation. As was mentioned above, each construction firm is unique in several ways. The number of possibilities is exponential, thus representing a challenge to design and develop a particular company. In most cases, new processes need to be tailored further by the individual company and its employees.

Our observation is that industry practice rewards the development of practical and short-term innovations more than seminal research. Companies in other industries such as aerospace, computing and engineering have budgets for robust and long-term R&D. Construction appears to benefit significantly from *developing* applications of others' breakthroughs to the sector and the investment pattern follows. Funds for research and development (separately from each other) are invested unequally in the AEC industry.

Reviewing possible solutions to increasing innovation activity, reviewing Kuhn & Schlegel's (1963) work might help. A structured approach is the only way to have multiple creative births or breakthroughs. Furthermore, there are mature and immature fields of research. Mature industries have much cleaned up after investigation is completed since they have deep research with disparate conclusions.

What is the answer? As a beginning point, our answer is simply a two-fold response. Firstly, we must realise that an innovative environment needs to be created for construction. This will facilitate "multiple births" by many entrepreneurs. That is, that all resources and focuses – public and private – must be brought to bear. Resources should include stronger relations between the academy, government and industry. As part of this, it is imperative to have contractor-directed research that is partially or fully funded by the industry. Innovation centres can be a hub for bringing together the practical with the theoretical.

Secondly, encouraging higher patent activity should be a focus. They reward not only breakthroughs but also significant improvements to existing patents. 20-year protection gives strong incentives, including financial rewards for those who can create solutions to our industry problems.

Construction's Risk–Reward Curve

Risk versus reward is an elemental concept of economics. It states that the risk-taker should seek an additional unit(s) of compensation for every increased unit of risk. In most cases, there should be at least a ratio of one to one. However, many times, it should be more units of reward. For example, this compensates for the high risk in the construction industry. In other words, if a risk is assessed as a 50% probability of loss, then to rationally price this risk, the minimum proposal to the client should be two times our total cost of involvement (plus a profit margin).

Conversely, if we take a minor risk then, we should not expect an enormous reward. Our expectation should be for a relatively small one. The payoff has a high probability of occurring. Australian Treasury Bonds are a typical example. They offer what has been labelled the risk-free rate.

Considering all possible project risks, their severity and probability can be overwhelming to construction contractors, subcontractors and construction service buyers. Additionally, organisational ones are significant. There are many uncontrollable factors in this industry. As a result, some universities have added a risk management subject to their construction management curriculum in recent years. This has been at the industry's suggestion, more specifically, industry advisory boards and alumni groups.

The relationship between risk and reward can be graphically represented as a line on an X–Y graph. Drawing this line helps explain its behaviour more clearly to all. Economists call this representative graphic a Risk–Reward Curve.

Construction's risk and reward relationship can be positive or negative depending on the context. A construction industry example of a positive relationship between risk and reward is one many construction professionals have experienced. A recruiting firm contacts you to see if you would be interested in a position with another firm. After an initial description of the job opening, its location and questions about your qualifications, the recruiter will ask a fundamental question, "What do you earn currently?" If your answer reflects a general business risk-reward curve, you might state, "It does not matter. You are asking me to take a large risk by going to a job where I don't have a professional network and the well-earned trust of my boss. If I go there, things may change as they often do in this industry. Construction is a project-by-project business, so I might be released after the project is completed. The risk is high; therefore, my salary should be $_____. It has no relation to what I make now, nor should it".

A well-accepted and straightforward example of a general business Risk–Reward Curve is shown in Figure 2.3.

Many organisational risks may follow Figure 2.3's linear characteristic. Examples are hiring an unproven but loyal person or implementing an innovative software and hardware package. Each action possesses a significant risk but a high reward.

Contrarily, we suggest that project risk-reward in construction contracting is not like the general business risk-reward curve. Instead, our risk curve is generally opposite to one of general business. That is, for every unit of raw risk, we receive a smaller reward. Conversely, for minimising raw risk, we receive a larger reward. Throughout dozens of jobs, our research suggests that this is valid.

To say it another way, there are identifiable potential events that, if they occur, result in an adverse financial event. Many times, when this happens, our cash-to-cash cycle slows down and expense increases (to pay for the problem). A contractor's cash to cash cycle is directly linked to their return on investment (ROI). Lower profit also affects ROI.

The reward or ROI we are calculating is after the project is completed. Therefore, it is affected. by risk events such as late payments, including retainage, loss due to claims, unpaid change orders, schedule delays, employee problems and other real-world issues.

ROI is the result of several financial inputs, wherein the two most important are cash flow and profit margin. To roughly calculate, divide the net profit dollars by the contractor's average cash invested in the project. Generally speaking, less risky projects have better margins and cash flow, raising the resulting ROI. (Figure 2.4).

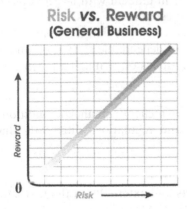

Figure 2.3 General Business' Conceptual Risk Reward Curve.

Construction's Risk vs. Reward

Figure 2.4 Construction Contracting's Conceptual Risk–Reward Curve.

Operating a construction business is not a negligible risk but a significant one. The industry is the second riskiest regarding the percentage of business failures in the United States. So, contractors are rational to ask for a substantial reward in return for their efforts.

The number of competitors is a significant barrier to asking for a fitting reward when pricing. Each contractor proposing a project assesses risk differently, as exhibited in varied bids. Due to the nature of this competition, lower margins result—currently, contractor average net profit before tax averages approximately 3%.

It is important to remember that construction contracting is a variable cost business meaning well-managed contractors, large or small, can be profitable. The size of a project(s) does not matter. Larger projects may be a firm's first choice; however, mega construction jobs will decrease as a percentage of total projects are built in an economic recession.

Our observation has been that small projects start closer to their planned start dates, have fewer stakeholders involved and have fewer payment retention requirements than more significant projects. There are other differences. In sum, these add up to less risk and a higher ROI. Whether exclusively or in tandem with larger projects, building smaller work is rewarded in strong or weak economies.

Before we start our discussion of risk factors, we need to state that it is assumed the contactor has identified their best type of work, has a dependable staff and is financially stable. From this assumption, a contractor can focus on managing project risk factors. Common risk factors include:

1   Longer project time – such as weather or change in client's financial condition.
2   A governmental agency that interprets codes and regulations differently than expected.
3   Change of laws and regulations.
4   Material availability and pricing.
5   Labour availability – internally between projects and externally in the market.
6   Project supervisor(s) quitting during construction.

7   Large project size – tighter margins, a more significant commitment of working capital.

8   Unknown clients – unpredictable as to norms expected.

So familiar people, processes and structures give construction firms more predictability of dynamics and less variability of results. As stated, static factors should be welcomed since there is little room for error between profit and loss.

There are several others. As a side note, notice that sales and marketing are included. We believe these processes are largely statistical. Additionally, our experience has been that a problem starting in early project phases many times will continue to affect closeout and final release of retention negatively.

These processes will drive you to your *sweet spot* on the Risk–Reward curve. This place is optimum for you and is always superior to the average return on investment. But, of course, your location on the curve will never be risk-free.

The value of this quantitative process is that it saves years of trial and error. Additionally, it starts discussions from a factual and unemotional place where issues are viewed from a purely business perspective.

To review, all construction projects have risk; however, less of these events in number, severity and time translates into a faster cash-to-cash cycle and higher profit margin, raising ROI.

For any contractor, many experiences in working with repeat clients in familiar locations make projects less risky. The risk factors do not change. However, the contractor's reaction to them does. They have found ways to reduce risk after experiencing each adverse event. This learned knowledge is one way, although inefficient, to manage risk.

Proactive management is critical. Conceptually, a contractor who systemises risk management quickly identifies and acts on early warning signs. A reduced risk project has improved cash flow and profit margin, which produces increased ROI. Higher ROI facilitates an increased investment for strategic improvements such as training, internal infrastructure or higher credentials to attract better clients.

The Supply–Demand Curve of Construction Contracting

Understanding the supply–demand curve is a critical step to creating and executing an effective strategic plan. Some academic models present a clear picture of economic realities. They are used to explain real-world behaviours. Tversky and Kahneman assert that theories order knowledge and predict outcomes (Lewis 2017). In strategic planning, it is essential to grasp construction business dynamics first before creating an effective strategy. Using a well-accepted model to illustrate how markets and clients behave is helpful. In construction, a visual representation assists understanding since its professionals appear to be more visually attuned. One of those models is the supply–demand curve of construction contracting.

Let us Define it First

The principle of the supply–demand curve illustrates how changes in supply and demand affect the price. The supply–demand curve for construction services portrays the problems and opportunities in the industry. It shows how good and poor construction can be as a market and a profession. Before we start, it is vital to make clear a couple of essential vocabulary items:

Price is the Client's Cost

Our price is the client's cost. In other words, when contracting prices are high, we mean that the market sees a high cost to build. Conversely, contractors have dropped their prices (and presumably decreased profit margins).

Supply is the Number of Construction Organisations

For simplicity purposes, we will keep our discussion contained to this point. However, please understand that the number of contractors drives capacity. Therefore, a decreasing population of contractors decreases the amount of construction that can be installed.

We will not discuss project prices, level, or future pricing movements. Instead, we will discuss how demand and supply affect project pricing. When we state the word cost, it is not the cost of material or labour to the contractor and it means the price of our construction services to the market.

The purpose of this section is to take a macro or global view of the industry. For business and strategic planning, this is a critical item. It answers the question: what will I do if demand changes? What is our plan "B"? Alternatively, how does one respond to changes in price and thus, affect profits?

The supply curve of construction firms is flat in the short term. The supply (or number) of contractors does not significantly increase due to a price increase. Furthermore, it does not decrease due to a price drop. In a robust economy, demand is still sensitive but less so than in a recession (see Figure 2.5).

Why?

People who are contractors will stay in the construction business. They might make their business smaller but will still own a construction firm. Contracting is a long-term

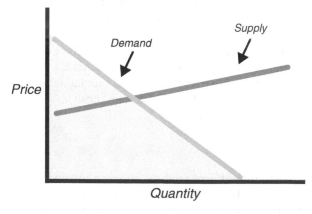

Figure 2.5 Robust Economy: quantity versus price.

skill that entrants cannot learn quickly, unlike real estate and have successful transactions. Although real estate is not an easy profession, it is widely understood. Most
people in real estate have owned property or a house before they enter the business.
Additional evidence is the relatively short training and qualification period when
compared to construction contracting.

Contractor supply does not rise or fall due to price (whether low or high-profit
margins) in the short term due to the following reasons:

Highly technical.

High risk.

Professionally unattractive industry.

Competent contractors and craftspeople do not increase quickly.

Most construction company owners do not desire another site of work.

Said another way, the short-term supply of contractors is inflexible.

Furthermore, as we look at the Australian construction industry from the early 1960s
to the present, we see the following:

The constant dollar value of construction is relatively unchanged.
The number of contractors has increased threefold.
To repeat, the number of contractors has tripled while the volume has stayed the
same resulting in the following effects:

Competition has increased
Margins have decreased.

According to ABS statistics, ours is Australia's second riskiest industry as a % of
business failures. Additionally, it is interesting to note that construction contractors
before tax net profit percentages have continuously been single digits regardless of
trade. (Figure 2.6).

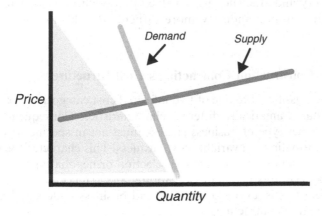

Figure 2.6 Recessional Economy: quantity versus price.

Quickly lessening demand during the increased cost means people have options other than building a new office, factory or home. In the short term,

> Do nothing and wait.
> Stay in the existing facility and, if necessary, renovate or repair what is essential.
> Build with in-house labour.

In the meantime, "bid shopping" to all contractors – qualified and unqualified – may occur. Overly focusing on cost allows weakly skilled contractors to sell their services aggressively, sometimes resulting in a project agreement. Those construction service buyers have regretted this.

Do not forget that building a new factory or corporate office is a top three investment for a business. On the consumer level, the construction of a home is the largest individual purchase a person will make. These decisions are not impulsive. The cost/benefit is weighed carefully and slowly, sometimes disproportionately focused on cost.

In summary, when:

> Demand goes down.
> The supply of contractors is mostly unchanged.
> Demand goes up.
> The supply of contractors is mostly unchanged.

We conclude that construction prices will never be driven up wildly, unlike sectors such as gold or internet stocks. People have options to building, including delaying, minimising or doing it themselves.

For the contractor, there are no windfall profits on the horizon regardless of the economy. This means they must be highly focused on skill-aligned work and disciplined in pricing for that work. Add to that, efficiency and quality is needed to realise consistent profits. The good news is that qualified contractors are limited in number. Also, contractors can decide to "go small" and still be profitable.

Most qualified contractors and savvy construction service buyers understand the above well. As shown, both know the market for construction services will never be generous, but quality and efficiency have value. Those who do not know these points make the second riskiest industry more difficult for themselves and others.

The Great Misconception: Construction Contracting's Cost Structure

Every type of business can be classified into one of two informal cost categories: Fixed or Variable. Each of these has dramatically different characteristics. Consequently, those who want to manage either type of business success must act in specific ways. For example, construction contracting is a variable cost business. This characteristic is essential to understand in planning for growth, project selection or negotiating with a client. Not to realise this may lead to some poor operational and financial decisions. Our observation has been that this is a critical strategic and business understanding one must have about construction contracting.

For each term, here are our informal definitions:

A fixed cost type of business is one when a sale occurs and costs increase by less than 50%.

Example organisations are airlines, computer software developers and restaurants, just to name a few. See Figure 2.7. In these same industries, costs typically increase by less than 25%. The sale represents a chance to make more than 50% profit if it occurs with other sales. Selling many pizzas or airline seats in a month and that month should be highly profitable. Fixed cost businesses drive turnover volume to make the economics work. This means their price to the client can be highly variable. To pay for this high percentage of fixed cost, they must sell a relatively high volume. In other words, most of their cost is not caused by a sale.

A variable cost type of business is when a sale transacts; costs increase by 50% or more.

Typically, these are service businesses such as construction contracting. Most of the business cost characterises these companies is caused by a sale. This is because they have minimal fixed assets. To say it alternatively, the monthly business cost reduces dramatically if there are no contracts won.

Another way to understand fixed and variable costs is to observe the contribution margin.

This is the gross profit that results once direct or project costs are deducted from revenue. In a fixed cost business, the contribution margin is over 50%, sometimes approaching 90%. In a variable cost business, the contribution margin is less than 50% and in construction contracts, it approaches 10%. (Figure 2.8).

We have yet to meet an owner of a construction firm who does not understand the variable cost nature of contracting; however, hundreds of employees, both site and office, do not. These employees are the same people who make dozens of daily operational decisions and participate in strategic planning. Therefore, they need to

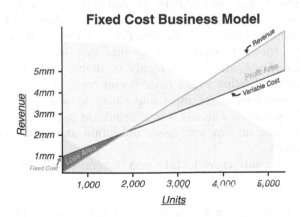

Figure 2.7 The Fixed Cost Dynamic.

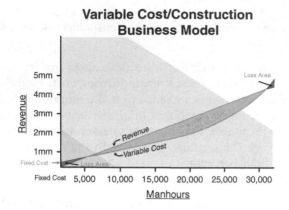

Figure 2.8 The Variable Cost Curve.

understand the business the same as the owner. If so, improved operational decisions and strategic inputs should follow.

The variable cost structure of construction is unusual, along with many other characteristics; the combination makes our industry unique. In contrast, many other businesses that serve and are served by our industry have a fixed cost structure.

Why do some contractors seek high volume? Sometimes, they feel outside pressure to grow their business. It is an interesting exercise to list firm types selling to construction contractors. This list might include:

> Computer Software Companies.
> Real Estate Professionals.
> Insurance Firms.
> Material Distributors.
> Rental Equipment Companies.
> Commercial or Residential Developers.

To illustrate, take two organisations 1) wholesale distributors and 2) developers. Each has most of its cost in fixed assets. Wholesalers have inventory, warehouses and trucks, while developers' costs comprise land and buildings. Both have salaried staff.

Each fixed cost business wants robust turnover growth since they have already purchased assets and need increased utilisation to be highly profitable. Said differently, more sales gives greater asset utilisation which reduces unit cost.

It is essential to state that labour cost is the most unpredictable of any input; however, it is a minimal part of their cost, whereas it is a significant part of a contractor's. Therefore, most contractors do not seek asset utilisation and they seek labour and management productivity.

Robust turnover growth requires contractors to take one, several or all the following actions:

1   Bid more projects, including ones that are not in the contractor's "sweet spot".
2   Propose larger projects.

3   Price to more clients, some that may be unfamiliar.
4   Hire more employees, many of whom are unproven.
5   Work with more unfamiliar regulatory jurisdictions.
6   Purchase from and logistically coordinate with unfamiliar suppliers (in these different jurisdictions).
7   Increase internal company infrastructure (such as offices, vehicles, computer hardware and software).

Each of these increases inefficiencies. Construction contracting is a single-digit net profit percentage before tax business, so the buffer between gain and loss is thin. As many know, the industry dynamics punish inefficient contractors.

Pricing to the client cannot vary widely in a variable cost business. There can be no "half-price sales event". This is partially due to the high number of competitors in construction, but also the contribution margin.

Since price variability compression is the norm, this places a value on seeking and winning any work and projects a contractor can perform productively, ones that are skill-aligned. In other words, to be profitable, construction firms must insist on bidding work they can execute well with a client or other stakeholders who act in the same way assumed when the cost was estimated.

The industry does not reward turning our business into a fixed cost one. For example, if a utility contractor buys (instead of lease/rent) equipment early in their first year of business, it will have a fixed cost or "nut" to cover. This "nut" is not going away for some time. The cost of payments is every month. This places more pressure to sell another job regardless of profit margin. On the other hand, it raises the breakeven turnover each month if breakeven costs are low, the ability to walk away from an unfair or overly demanding negotiation.

There are several ways to transform a construction firm into a fixed cost business. Here are examples:

1   Higher Salaries than the Market Average.
2   No Incentive Compensation.
3   Higher than peer average Overhead to Direct Cost Ratio.
4   Asset Purchase or Long-Term Lease of an Asset with Unknown Utilisation.
5   Receivables Allowed to Lag.
6   No Retained Earnings.

It is crucial to address any early indicators of these financial symptoms.

Be aware, site and office staff can be a "de facto" fixed cost if a contractor is unwilling or unable to reduce headcount. Just as buying a backhoe on a payment plan, employing dozens of people when the project pipeline is low puts pressure to sell another job. With an unconditional commitment to people, there must be a consistent stream of revenue. Otherwise, layoffs must occur to keep costs in line.

To be fair, contractors will keep people employed until there is no other alternative. However, they know that once a reduction in force occurs, the odds are high that many will not come back to work. This is one of many reasons that construction firms are hesitant to adjust their employee's headcount.

One of the keys to the variable cost model is that *break-even* (the turnover needed to cover all costs) can be achieved with a minimal amount of revenue.

Most contractors do not have significant recurring or operating costs outside of job costs. Construction firms can work out of a modest office or even their home. This expense is minimal. Furthermore, their profit goal can be a living salary. Not a significant amount of volume is required to produce this. It is one of the benefits of variable cost dynamics.

This gives contractors flexibility in finding the right kind of work. They do not need to sell every customer, just a few who pay well at a proposed price that includes a reasonable profit. This is evidenced by "hit rates" of financially successful contractors of less than 20%. Producing many accurate estimates to clients allows you to charge a little higher price but a proportionally greater net profit. Said another way, financially substantial construction firms have a return on investment in projects over 30% due to being choosy of which projects and clients they engage.

We should also mention that this extra margin gives a contractor a cushion to pay for unforeseen and uncompensated costs. Construction is risky. If a project is problematic, a construction firm has added margin to buy extra labour and material elements.

History shows that if a contractor increases their turnover too quickly, costs rise, and profit margins shrink. Restated differently, expenses and profit will cross over at the high end of the cost/turnover curve.

This is due to two factors:

The last 50% of the volume must be sold cheaper than the first 50% of the projects. As a result, projects must be negotiated at a discount to attract more contracts, making the turnover curve slightly downward.

Adding extra, unproven people make a firm inefficient. Mistakes, rework and a general slowing of operational processes occur. Also, you may be selling to unknown clients in unfamiliar jurisdictions. These raise the cost curve.

Again, the cost and turnover curves will cross the top end, spelling financial trouble. Higher volume is rarely the answer. Finding the right amount is critical for above-average profits.

It is no secret that the best approach to profitable contracting is to be financially conservative. To be clear, our working definition of financial conservatism contains three points:

Being frugal in acquiring fixed assets.

Keeping site and office efficiency high.

Charging market rates for work.

The first point is evident to all business owners. This is known as "squeezing a nickel", i.e., negotiating the least cost when purchasing products or services. Companies that practice this will usually have a cost advantage over most competitors.

However, points two and three are sometimes forgotten. True financial conservatism or not buying cheap labour or discounting the market price to clients-it is about insisting on efficiency and proposing market-level pricing and terms. These three tenants are basic to earn greater profits.

Remember that many things in the construction business are different from other industries. The risk–reward curve and supply/ demand dynamics are good examples. Knowing these foundational characteristics make for a robust strategic plan, i.e., realistic and insightful. Additionally, being aware of these differences makes any contractor less prone to operational mistakes and, thus, consistently profitable.

Do You Have a Continuous and Balanced Improvement Culture?

Experienced contractors know that the construction industry's constant change requires each of us to pursue continuous improvement. It is a choice and controllable action. Its benefits can give strong strategic and operational results long-term. Sometimes, innovation is prompted by a problem, internal change, industry trend, or innovation. A bad idea should not be rejected immediately. Instead, the true innovator tries to find where it fits (Lewis 2017). In each case, a strong internal culture is ready to act and support the journey to increasing efficacy making it normal and expected. Dauntingly, there are many issues to address before continuous improvement can be effectively pursued in a construction firm. See Tables 2.2 through 2.4.

Continuous improvement is complicated in a construction firm. The journey is difficult, such as atypical projects, complex material logistics, unpredictable employee conscientiousness and unique contracts. More critically, most contractors are from the Boomer generation and many managers are from Generation X, while new entrants are Millennials. Contractors can be conservative business leaders and trust tried-and-true methods. Therefore, they are still in business when their competitors have failed. However, there are innovative ways to construct that the competition is pursuing and will master. In our experience, high return-on-investment (ROI) companies have cultures that are not first to innovate but first to adopt a proven idea.

Table 2.2 Comparison of Lean and Six Sigma Improvement Methodologies

Characteristic	Kaizen (Lean)	Six Sigma	Comments
Years in Use	Post-World War II	Since the 1970s	Decades of business use by private firms mean these methodologies deliver value
Notable Practitioner	Toyota	Motorola	Both are manufacturers
Major Focus	Culture Change	Project by Project	Construction firms have used both
In a Phrase	"Change for the Better"	"Eliminate Product Defects"	Six Sigma reduces the defect rate to 3.4 per million – not a reality in construction
Scope	From Lowest to Highest Levels of Company	Executives Prompt Problem-Solving	People who do the work are the experts
Problem Solving Focus	Eliminates Waste	Less Variability	Both issues are prominent challenges in construction
Process	Plan-Do-Analyse-Change	Define-Measure-Analyse-Improve-Control	Both appear to fit construction's business model

We assert that the most successful companies have mastered three balances:

1   Between innovation and proven methods.
2   Between rapid and incremental improvement.
3   Between highly detailed documentation and trusting employees for part of the decision making and execution.

A CBIC is a subset of a Generative Culture. In essence, a Generative Culture is one where everyone behaves like an owner; they feel its success is their success.

In a CBIC, employees are self-critical and have a "compete against myself" attitude. Employees from the CEO to front-line workers strive to find new efficiencies and create new opportunities to generate profit. They understand the long-term power of "making your best better" – it affords many professional and personal options.

Continuous improvement is for those contractors who play the long game. Those who see their business as getting more valuable over time and that being focused on long-term value for the client and the contractor is the best focus. It is not a quick fix or a weekend seminar but a lifestyle.

Continuous Improvement is an old concept as relevant today when it was first conceived. Improving systems is well-known as a universal challenge among for-profit companies, including construction contractors. Yet, there are few formal improvement programs available for review and implementation. The two most prominent are Kaizen (A Lean Concept) and Six Sigma. To familiarise ourselves, we list the major characteristics of each in Table 2.2.

A Quick Look at Construction Innovation History

From our experience, the answer appears to be both. First, problems may be the motivation needed for innovative thinking. This is a variation of "necessity is the mother of invention", which is alive in the construction industry. Second, the innovation process occurs when one person says or does something and another person understands it more thoroughly and extends it (Lewis 2017).

The "triangle of knowledge" comprises research, education and innovation (Abramo et al. 2009). We believe experience and observation must be added to create a truly creative and implementable idea for the construction professional.

However, non-construction innovations may also be prompts to apply those new products or ideas to the industry. Here is a partial list:

1970s: Computers – Computers were an outcome of the Military's and NASA's efforts to calculate numbers non-manually and faster. In addition, national defence and space exploration were government focuses thus, well-funded so they could survive failed experiments.
1980's: Partnering – The creation of many professionals, including the Construction Users Round Table (CURT), that observed delays and resulting litigation were becoming more common, expensive and chaotic. Partnering saw the wisdom of planning for conflict while keeping project relationships strong.
1990s: Safety – The emergence of an ethically important issue prompted by more recording and analysis of workplace safety. It was formally organised as a requirement under the penalty of law.

2000 to present: Handheld Electronics – The power of information delivery and decision making, in one hand. This emerged as a natural consequence of computing improvement and communication technology. It was not from a problem, but as "a solution looking for problems".

Pre-fabrication – Construction components constructed offsite in parallel with other related tasks onsite. Although identified as a possible improvement over a century previously, significant private investment emerged to accelerate the development of approaches and systems resulting in a noteworthy built environment.

Drones – Less "managing by walking around" (MBWA) making travel to observe projects observations non-existent for executives and managers.

3-D printing – The power of technology trying to solve the labour shortage, productivity and quality control challenges.

It is interesting that innovation adoption typically requires 18 months for full implementation. Indeed, after this initial phase, improvements and new thinking are needed to keep the Innovation current and a competitive edge over other contractors who adopt the same system.

Regardless of the impetus for change, it is essential to remember that taking on recent technologies is not necessarily an indication of a CBIC. In our experience, CBICs result from teaching, mentoring, a steady company process and a purposeful mission.

Addressing Construction's Characteristics: The Efficient Track to a CIBC

One challenge to developing a CBIC is lowering employee turnover. Construction's unique nature erects barriers to Continuous Improvement and one of those barriers is the workforce change rate. There is a remarkably high employee turnover ratio in our industry. The latest governmental data show that the turnover of hires, re-hires and separations is twice as much as manufacturing. Creating and executing an improvement process supported by a robust organisational culture must be tailored to this unique set of industry norms. See Table 2.3.

The Case for Incremental Improvement

As stated, we recommend a culture that supports incremental improvement. There are many instances that construction firms have undertaken rapid and multi-practice improvement programs. However, we know that these efforts have not brought the anticipated change hoped. Construction contracting employees are wary of rapid change as is the industry.

We have seen more success when trustworthy practices are standardised and incremental innovations are made. Additionally, in some companies – as strange as it may sound – getting every employee to follow existing processes is a challenge. This is partially due to the transient nature and previous learning of construction people. Some may say stress discourages them from changing their approach and this coupled with the confidence of a tried-and-true method erects a barrier to improvement.

Contractors must decide their pace of implementation: many (MPI) or singular (SPI)? Our time frame is on a per quarter (three-month) basis. We compare the efficacy of these two approaches (see Table 2.4).

Table 2.3 Issues to Address in a Continuous Improvement Initiative

Characteristic of Construction Contracting	Potential Problem Issue Poses	Potential Solution	Comments
Employee turnover percentage is approximately twice other industries	Training expense of new and re-hired employees while those leaving are taking that investment with them	Employee turnover can be lessened by committing to a consistent or slightly increasing turnover amount	Eliminate or reduce this significant percentage
Fast business formation means hyper-competition	Company perception is that rapid innovation is needed to stay ahead of a high percentage of new entrants	Share government statistics that show business formation	Approximately 6% per year over the previous four years
Single Digit % Net Profit Before Tax	Fewer dollars for correcting mistakes	A formalised process to identify improvements	Small projects division may be the best place to start
Craftworker's unique process learned over her/his lifetime	The craftworker's approach is well embedded and has been successful for her/him	Focus on the needed inputs to the craftworker such as material logistics or information delivery	Technical craft improvement methods will be difficult
Unique design and construction challenges for each project	Cannot make a "Detailed System" for every contingency	Planning, communication and risk management are primary focuses	A process map is focused on few streams of practices

Table 2.4 Comparison of Multi versus Singular Practice Implementation

Factor	Multi-Practice Implementation (MPI) Per Calendar Quarter	Singular Practice Implementation (SPI) Per Calendar Quarter
Training	Much training needed consuming time and money	Single Practice training can be done "on-the-job"
Late Adopters	These employees are a large barrier since many practices are implemented during a quarter	Late adopters will be less of a problem to manage
Construction's High Employee Turnover	New people are often a barrier to rapid and complete adoption	The knowledge gap is less for new employees (since it is a single practice) as compared to MPI
Industry's 4% net profit average	Many practices implemented might push the firm into negative profit results	A one-practice-per-quarter change threatens profit margins less
High-Risk Industry	Raises chance of adverse Safety, Quality and Financial outcomes than SPI	Less risk than MPI approach
Full Implementation and Compliance	Less of a chance than SPI – Much change means "overwhelm"	More of a chance than MPI – Balanced change means deeper focus
Possible Conflicting Practices	High chance of conflict resulting in confusion	No conflict since 1 per quarter is adopted

Extending this thinking, the legendary quality expert Joseph Juran (a contemporary of and collaborator with Deming) pointed out that companies must target areas to innovate and designate areas to keep static, so the improvement focus is high and purposeful. In addition, designating static areas helps keep the complexity of innovation lower.

As we have asserted, too much change can be problematic. FMI's recent study lists root causes for tier 1 and 2 contractor failure. Interestingly, each of these is controllable: 1) Too much change – 90% 2) Poor strategic leadership – 76% 3) Excessive ego – 62% 4) Inadequate capitalisation – 58% 5) Loss of Discipline – 45% (FMI Quarterly, 2016) http://www.fmiquarterly.com/index.php/2016/06/14/why-large-contractors-fail-a-fresh-perspective/. Our experience tells us that a culture supporting incremental change addresses these problems. Cultures are shared beliefs. The best ones have leadership that relies on logic and facts to establish those beliefs. Team cultures self-check and therefore, members are free to pointw to group dynamics such as weak leadership, big egos and unsubstantiated assumptions.

Highly Detailed Documentation is Inefficient – Except for Safety

We know of instances where contractors have engaged consultants to document all non-safety processes in detail (while sometimes instituting a non-construction vocabulary). The result was frustration between the consultant and contractor. To put it in technical terms, there is too much variation of inputs to capture, along with a steep conceptual learning curve by a non-construction consultant. Construction is not manufacturing nor a "commoditized" service. Adverse events often occur with muted warnings such as project delays, logistical problems and scope changes. Late project delivery triggers significant penalties. Add to that many times, outsiders have made mistakes (client, designer or supplier), but no timeouts can assess and sanction. The constructor must notify, but often "soldier on", then sometimes wait until the end of the project to see if extra compensation may be decided in their favour.

We find that a well-structured way to document individual non-safety practices is to use prompts such as *Steps, Deadlines and Cardinal Sins*. These give "guardrails" to methods in such areas as planning, communication, execution, measurement and adjustment. Additionally, to improve individual practices which make up overall processes, capturing the existing ones with clarity allows for improvement to be suggested, contemplated, then, if agreed, adopted.

In summary, many times, construction's highly variable project inputs and resulting problems should sometimes be addressed by "it depends". This means that, based on experienced people interpreting complex situations, the project team knows what to do and not do. But, unfortunately, this kind of stochastic answer can neither be mapped, narrated, nor storyboarded.

We assert that balance in contracting is critical since the penalty for mistakes is high. Balancing three choices in pursuing continuous improvement are between 1) innovative or proven methods, 2) incremental or rapid innovation and 3) highly detailed documentation or trusting the employees. So, continuous improvement presents three competing realities. The proper balances make for better improvement quality, "stickiness" and outcomes.

A CBIC supports today's journey, which will evolve into future opportunities. To increase effectiveness without rework, a structured approach is needed. Rework is the most expensive of all mistakes. A disciplined pathway will provide efficiency. Much time and energy can be spent planning and executing an ineffective idea. We suggest structuring your improvement process with a framework, then documenting and communicating it to capture improvements, provide clarity and establish a platform from which others can build upon for the next generation of leaders.

Contractors and their people who improve over time have increased options who do not. As company owners retire and exit their firm, a quality-minded and efficient contracting organisation has a value that can be sold to an outside buyer or transferred internally to those who want to carry on the enterprise's purpose. Employees will have options as the quality of their work and thinking will attract attention from executives who know these savvy people can lead safer, faster and less costly operations.

Construction's Overhead Efficiency

All formally organised construction firms have a home office. The cost of its operation is typically referred to as Overhead. Generally, home offices support project operations and acquire new work. One major indicator of the profitability of a construction company is how well a company leverages its overhead to build work safer, faster and better. Those firms that have a top-quartile efficiency (low overhead expense per project dollar), often have top-quartile net profit before tax. We believe that the critical battle to be won or lost is at the job site, where the direct cost of constructing comprises approximately 80% or more of contract revenue. However, efficiency at the job site can either be supported or undermined by the office. The people, practices and company infrastructure that constitute overhead often translate to efficiency – or inefficiency – onsite operation and work acquisition.

Definitions

Before we proceed to explain the issues and potential answers involved, we need to define terms.

Overhead Cost – cost expended in the management and administration of the installation and rehabilitation processes of construction.

Direct Cost – funds expended for site employees, materials, equipment and subcontractors to directly construct, rehabilitate and upgrade of shelter, processing facilities and infrastructure.

Non-Production Employees – executives, managers and office staff of construction firms. For our purposes, these types of employees are part of overhead and its cost.

Production Employees – craftspeople, equipment operators, journeypersons, labourers and others who directly work on projects. For our purposes, these types of employees are part of direct cost.

Yearly Average Employment – this number is the monthly average of persons employed by a construction firm for the year noted.

As a further clarification, we assume that all efficient contractors are highly profitable and vice versa. So, those terms will be used interchangeably in this section.

Improving Overhead Performance

We know that the firms that improve their overhead performance improve profitability. The question is how these successful firms make the best use of their general and administrative expenses. Overhead has evolved in the last decade. New management positions such as Risk Management, LEED, Pre-Construction Services, Virtual Design and Construction (VDC), Diversity and Recruiting have emerged. Some mature companies have established a Board of Directors. This has added to the number of office people and overhead expense. Some of these positions have been created to address new regulations, while others answer an industry reality. However, we have observed efficient contractors define these roles in ways that add value to the client and, many times, help construct work.

Pay More Attention to the Overhead-to-Direct Cost Ratio

We believe that overhead best indicates the efficiency of overhead to direct cost (OH/DC) metric. Whether markets are growing or contracting, the battle for profits starts with overhead efficiency in driving the direct cost to produce a bottom-line profit. To restate, if fewer overhead dollars are needed to install the same amount of work as your peers, then higher than median profits will follow.

Additionally, these two cost categories are the largest on contractor's profit and loss report. Minor variations due to unplanned purchases or events do not significantly change the resulting percentage. The added value of its comparison with other contractors nationally and regionally is substantial. This delivers a dose of reality to any firm that is especially needed in challenging times or overly critical leadership.

OH/DC is comparable to other peer contractors via the banking data. It is a compilation of source documents such as tax returns and financial statements. As you know, only construction firms that stay in business produce and submit this information and hardly anyone submits fictitious information. This data is credible.

To quote a colourful client, "half of the construction contractors are below average". We logically know the other half are above it. A straightforward comparison with underperforming and over-performing peers is the first step towards getting into the top half.

Some people may point to project variations that invalidate the OH/DC metric's value. Variations do not skew the ratio significantly. The overhead expense and direct cost are yearly accumulations of all jobs. Additionally, a project that has increased its contract amount due to variations will undoubtedly have some time extension. A company's overhead average "run rate" is hefty each day and moves somewhat proportionately with increased direct cost. Variations many times have higher profit percentages than the client contracted amount. However, the impact of extreme variations on a single project will not be significant as part of all projects, i.e., total direct cost.

We have observed that construction service buyers are motivated to prompt designers, funders and others to help contractors complete construction in solid economic times. There may be many reasons for this. We have observed in several situations:

When the real estate market has high demand, tenants pay more dollars per square foot to occupy a property. Accordingly, developers can start profits and cash flow earlier with a completed project.

Existing infrastructure is at capacity due to a burgeoning population. For example, city and state governments hear complaints from constituents about losing time and money due to slowing transit times and brownouts.

Public buildings delivered timely are a sign of governmental efficiency. Early completion is a dramatic example. Conversely, lateness is a sign of weakness as seen by voters.

With these points in mind, are there consistently helpful clients? For example, do they employ engineers and architects who are realistic in their design and timely in their responses? Can choosing a few more of those types of clients help your efficiency?

One metric sometimes used by contractors is overhead to revenue. We find this less exacting. In good markets and bad, contract amounts vary for the same project depending on economic time and thus, net profit margins fluctuate. The overhead cost per turnover dollar drops when margins are more considerable in good markets. In poor markets, the reverse is true. Overhead to turnover fluctuates more than OH/DC through no fault or savvy of the construction firm.

Also, for apartment and condominium builders who lease those units, how do they know how efficient they are? There is no turnover to book upon completion. Of course, they may look at the cost per square foot of finished construction. However, since the overhead cost of these builders hovers at 10%, these are significant dollars focused on constructing work.

Interestingly, emergency service contractors do not have any control over their revenue. Acts of God and human migration are the two leading causes of work. So how do you fairly apply overhead to turnover for these contractors? How efficient one is from the start of the disaster to the end is to measure how the office (OH) directs the site (DC) under high pressure and a short time frame.

Disturbingly, overhead for all contractors "creeps" over time. This is typical of all hierarchies. They tend towards a less merit-based system. For example, when a firm hires a professionally educated person, this person's ambition may be to grow their responsibilities, i.e., their career and compensation. Sometimes they want to add assistants to do more tasks. Other times, somebody might request technology licences and association memberships. In a bureaucratic construction firm, this may represent an "arms race" with other corporate staff, i.e., who has more perks and permissions? No one asks the confronting question, "How does this add value to the client or help build projects better". Remember, the construction economy eventually will slow and unnecessary expense is jettisoned. Efficient contractors do not wait for the next recession.

Improving OH/DC: Recommendations for Contractors

Now that you are thinking about OH/DC, we recommend you gear the firm's processes towards better monitoring and evaluating this critical metric. The goal is to lessen the variability of outcomes in timing (late), quality (poor) and cost (more dollars or increased effort).

Due to the many unique interactions between employees during project pursuit, construction and administrative/financial closeout, one mis-executed step can have multiple adverse ripple effects downstream. In an average performing construction firm, we have observed many late or incomplete practices that are still handed off to the next person, making for dozens of reworks and "workaround" events. To us, the list below is the straightest path to less overhead headcount and more direct cost throughput per dollar of overhead.

1   Create a list of activities and segregate them into three areas.

   a   Start doing.
   b   Keep doing.
   c   Stop doing.

2   Process maps your complete set of practices, from Work Acquisition to Project Operations to Financial Management along with Strategic Planning and Business Management. From step 1, actions to start doing and keep doing must reside in this map.
3   From the process map, operations and business manual should be written in laymen's terms with company practice definitions and steps, including those not to do.
4   This manual creates a task list and monitoring process to measure practices compliance – early, done once and completely.

The information needed to create 1–4 resides in the *total knowledge* of the firm – from the person with the least responsive to the most. Engaging everyone in every step creates documentation and active monitoring where there is a more objective judgment of an employee's performance.

Compliance with company practices is the quickest improvement path. Complying with processes reduces rework, keeps crews and equipment working and ensures materials are there when the site needs them. In addition, complying with good practices is demonstrated to work, so making this part of your company's everyday expectation involves no trial-and-error risk or change management. The strategic effects are many.

Here are a few:

1   There are few "dark corners" in the company – places where only one person knows how to perform a function. This is a reasonable risk management practice.
2   If executives consistently reward and penalise personnel based on compliance, the company culture will evolve where on-time task completion is a primary daily thought in employees' minds.
3   The documentation facilitates training and immerses all employees in core practices. In addition, every meeting can end with a review of a selected practice's requirements.
4   This documentation and monitoring process is purposeful. We have found that following steps 1–4, the company's human dynamics become clear. Clarity will emerge as employees execute their assigned functions well or not well. All employees have blind spots, gaps in knowledge and job duties they would instead not do. This leads to uneven compliance to the practices that can make a company a high performer if done consistently.

Indeed, a construction leader does not need to wait to document practices and map their workflow to discover inefficiencies. Some overhead inefficiency is apparent. We have often seen project managers travel to the site each week, rarely working at their office desk. Their paperwork timeliness and completeness were mediocre. To us, this signals their passion for the job site. Once this was realised they were placed on-site and were much happier and, predictably, more effective.

In another instance, a project manager-estimator showed the skills of a gifted teacher and mentor. After this had been identified, the person was given a direct leadership role of the entire work acquisition department. In both cases, people were shown by their passion for what they loved to do and efficiency followed.

Said differently, if a person is given two types of projects to build simultaneously, they will pay more attention to one over the other. They will choose to be better at one kind of work – the reason does not matter. This is natural.

Additionally, a construction leader who employs an overly emotional and dramatic manager causes some of their inefficiency. We have seen these kinds of people distract others with lengthy non-core discussions. In one case, a human resource manager's behaviour reflected employee advocacy primarily. This caused slow firing, careless hiring and personality-based promotion recommendations, along with less time focused on the main thing – constructing work.

Interestingly, the quietest employees often pleasantly surprise their leadership and the loudest ones disappoint. Timely compliance with company practices is the differentiator.

In summary, a construction firm's profitability is driven chiefly by one factor aside from winning profitable work. It is overhead efficacy in leading and managing direct costs. This is controllable. Having a well-organised, directed and inspired office staff who keep the site fully resourced while eliminating distractions has helped many construction firms turn around their fortunes.

For the striving construction firm, we suggest attaining the top quartile is achieved by multiple choices. Of course, it is a journey and, at times, a struggle. But, for obvious reasons, it can be a worthy operating goal. Moreover, competing against those already existing firms can be the start of an improvement initiative.

If we were asked to reduce our advice to one sentence, we would say, "The purpose of a home office to provide the right inputs to the site operations, clients, regulators and vendors at the right time". Of course, this might be fodder for an excellent brass plague posted in a home office entry.

In this era of changing expectations and realities, we believe that each addition of overhead should be thought through with a cost-benefit in mind. By honestly answering the question, "Will this specific role adds value as perceived by our clients or build our projects better", does a company keep overhead efficient. Sometimes, combining two roles reduces a company's OH/DC cost ratio. We might call this "reverse overhead creep".

Without exception, the firms we know aspire to reach or keep upper quartile efficiency – top 25% of OH/DC as compared to their peer competitors. This affords many strategic advantages, including high cost and schedule competitiveness. This is especially important when the economy inevitably slows. Moreover, regardless of the economy, achieving the top 25% OH/DC efficiency leads to top 25% of net profit before tax.

The First Best Practice for All Construction Contractors

Most construction contractors have a set of good practices that work for them. Many have not yet considered that there is a first *best practice* that, when implemented, delivers more consistent, more sustainable results across the industry. This first best practice is adherence with processes. These practices include good craft skills, which is the only method in successful construction projects.

Best practices have garnered much attention in construction over the last decade. However, the term is still a buzzword among many professionals in the industry. Construction contracting still does not have a definitive list of universally agreed best practices. Industry discussion is mainly driven by opinion; however, we see an opportunity to increase data acquisition and thus clarity.

One challenge in construction is to be more definitive about which practices sit in which bucket (best, good and poor techniques). However, this is not enough. Contractors must implement these practices consistently and diligently. Monitoring *adherence* to what works is the best way to improve the essentials in construction, timeliness and quality of effort. Construction firms that consistently adhere to best practices win more of the time.

Processes and Their Practices

Best practices can be classified into five sets of processes that any successful construction contracting firm must do: 1) Work Acquisition, 2) Installing Work, 3) Financial Management, 4) Business Management and 5) Human Resource Management. Each is self-explanatory. From these five overall processes, there are hundreds of practices.

These practices range from small, daily decisions like which supplier to pay first to large, significant decisions like staying small or growing strategically. To define broadly, practices are those acts that can be viewed, measured or inspected. Either through file inspection, work review or observation, these proactive actions are focused on producing a result. Each practice done well affects safety, quality and productivity. These processes, robust or weak, are executed each day.

The company owner does not execute most of the construction contracting practices in our research and testing. Instead, they are performedby company staff, from site labourers to middle management. We have studied these staff, examining how they adhere to best practices across successful and unsuccessful firms. For 154 practices and across dozens of companies, we have done this as follows:

1   Asked company staff must rate each practice's importance. This taps into insider knowledge that is formed over many experiences with the procedure. Generally, industry data shows that construction employees have worked for more than one firm. As a result, they have a helpful perspective.
2   Asked each staff member to rate compliance with each practice. Combined with importance, another important metric is compliance. How often an employee complies with factors he thinks are essential is a creditable measure if a statistically significant number of qualified participants are surveyed.

As we describe below, compliance generates on-time and on-budget outcomes.

Adherence to Practices

As we have stated, we believe the first best practice to consider and implement is compliance. This can be done in several ways or as a combination of these. Electronic measuring is preferred in many situations, although face-to-face meetings have their unique value.

   Compliance in executing practices is a straightforward concept. In our research, it has three components:

   Practice is performed correctly.

   Practice is executed only once (no rework).

   Practice is performed on or before the deadline.

Indeed, practices executed with these three attributes are performed slowly and carefully; however, once done, their outcome is given to the next person for their use in producing safety, quality and productivity. However, the question remains what the average compliance based on these three criteria is?

   One of our clients shared a recent set of data with us. The company furnished data of completed projects closed over three years in sequential order, measuring process compliance, cost and schedule results (See Figure 2.9). There are two takeaways from this graph.

   Figure 2.9 also shows a significant improvement over three years. For clarity, the first five and last five projects represent before and after adoption reasonably accurately. Higher adherence to efficient processes is correlated to improved outcomes. The first five's unweighted outcomes were average project profit decreased (−81.69%) while exceeding the schedule (18.92%) than initially planned from lower compliance to processes (29%). The last five project's profits improved (48.68%) over the baseline estimate while shortening the baseline schedule (−28.78%) with higher compliance to processes (92%).

An Example of a Generic Best Practice

We offer the following "Best Practice" as a detailed example of what we consider as such.

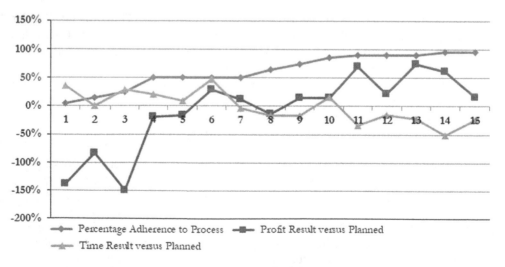

Figure 2.9 Results from Completed Projects over Three years.

From our library, one Best Practice example is *"Our job cost system allows managers to input completed units, compares them to estimated and forecasts cost at the end of the project".*

It is a simple practice and not one that is new. Indeed, the highest purpose of a job cost report is projecting cost to the end of the project (proactive), not what has been spent (reactive). However, to the building profession, this purpose may not be obvious. The power of this simple idea is that it has positive effects on multiple activity streams (Project Acquisition, Construction Projects, Tracking Projects and Organisational Management) in a construction contracting business. This practical outline we call *The Ripple Effects Framework* helps us determine Best Practices with clarity. Let us list these effects of the specific practice mentioned above. This practice of using completed units of work and using them to calculate % completion and forecasts cost-to-complete leads to:

1 A straightforward calculation of what billing is proper and earned by the contractor.
2 If a subcontractor to your firm is involved, their billing to you is clearer (and fairer) than using a percentage method. Since this is straightforward, there is less room for argument and thus, relationships should stay intact.
3 Places a healthy focus on estimating performance since estimated unit counts and unit costs are part of this system. The accuracy of the units estimated versus actual will show great competence or areas where improvement is needed.
4 The focus is on the completed work unit and not half-done installation percent completion guesses. The counting of completed units can be done confidently by a less-experienced field employee.
5 Produces a productivity measure (not production) per crew and projection pointing to either an estimating error (in unit cost or unit count) or field inefficiency.
6 Since the focus is on installed units correctly, it forces the discussion about what is completed by the project's quality standards (Specifications and QA/QC).
7 This counting method may also be used for client billing – counted once, but used four times – our projected cost, subcontractor billing, schedule update and client invoicing.
8 For those contractors who construct repetitive type structures – horizontally and vertically – this leads to linear scheduling – a simplified but accurate method that makes scheduling practical and more manageable.
9 A much easier over/under billing calculation and cost-to-complete for the company, including the CFO.

Nine positive effects, in our experience, is an extraordinarily high number of reasons – some compelling, to change an existing approach. For those construction firms with a small projects division, it may be a good starting point for implementation.

Of course, it is always best to statistically test a group of practices that represent an entire construction operation. We have done so, starting with Stevens's PhD dissertation and continuing that work to the present. However, we know from this study that there are few universal ones. Interestingly, of the 154 we tested with nine distinct types of contractors, only seven tested best.

Once a framework is confirmed, does this assist us in determining "Next" Practices (future Best Practices)? More precisely, what practices will be needed to answer emerging industry conditions? We answer this question with the framework outlined above. It can be used to test promising ideas. If a practice positively affects many areas of our company, there is a high probability of successful Next Practices. To restate, we should ask, "What is the effect on the four streams of tasks – Acquiring Projects, Constructing Projects, Tracking Projects and Managing the Organisation?"

Choosing the Next Practices

Selecting the "next" from the "rest" of the practices available is a decision that is highly specific to a firm's craft, culture and people.

Discussion and group scoring of the Next Practices' effects is a healthy exercise if a company implements consistently future practices that have positive impacts.

The generation of potential Best Practices can be from many sources. For example, the same list we cited above: 1) Empirical study 2) Use by a respected firm 3) Benchmarking 4) Survey, 5) Lessons learned and 6) Delphi Panel (a group of experts).

Most contractors have developed trusted sources from their interpersonal networks, including national conventions and association meetings. In addition, new ideas are the fodder of many construction industry speakers and consultants. However, the valuation framework for these potential Best Practices appears to be missing. We offer our Importance-Performance to Overhead/Direct Cost Statistical Test and our Ripple Effects Framework to fill this void.

Implementing Best Practices – One or Many?

Our experience and observations tell us that Construction Contractors go out of business more often by violating the basics of good work acquisition, exacting constructing of work and promptly collecting payment for that work than by not innovating their business. Thus, rapid innovation is a significant risk. But, again, the average net profit margin before tax is 3%. So, a construction business does not have a financial buffer to pay for the unintended consequences of rapid innovation. As you may know, slim margins are partially due to the industry's many providers – approximately 385,000 contractors. In summary, this extremely competitive industry lowers profit margins making rapid innovation risky.

In summary, there are fundamental flaws in many Best Practice Assessments. One of which is the determination process of a practice's value. Is data or opinion the basis? What was the root metric used to correlate its effect? How large is the data set and what is the span of trade sectors? When chosen well, correlations can be confidently assessed and conclusions made. We start the discussion about the most timeless metric with Overhead to Direct Cost compared to your peers (regionally or nationally).

Practices are subsets of processes. Practices are articulated briefly – 25 words or less – and understood quickly. As a result, their importance and performance are more measurable by those who execute them than a process articulated in a page-long description. This is a critical insight as contractors do not build alone but do so with the help of many people.

It appears that compliance to process is never considered as a Best Practice. From our data, compliance to tried and true practices by a firm is never part of the discussion. It should be. Our data confirms its value. Indeed, life's experience has taught us that consistently executing simple practices – such as arriving early, staying later, homework, note-taking and objective thinking facilitates success. Our data show that the most robust methods are known. Said differently, part of the solution to construction's productivity decline is in plain view.

When data is unavailable, we think another method must be used if a company is to improve without significant trial and error issues. We call this test – The Ripple Effects Framework. This framework is based on identifying positive outcomes in 4 streams of activity.

Best Practice is a loosely used term. Next Practice is an emerging concept that also represents value. We hope you better understand what these terms mean and how these concepts should apply. Also, we hope that we have given you a practical approach to determine which practices are best for your firm, now and in the future. We believe that a Best Practice enables an improvement in things that matter – safety, quality, cost and schedule. From these outcomes, benefits accrue, including top quartile productivity, to the construction firm which identifies and executes them.

Good ideas are many and many have been tried and failed. The construction industry is complicated and its contractors are unique. Each has its way of doing things. Indeed, each contractor knows what has worked best for them. They have significant incentives to know this. Those incentives include keeping their construction business alive and thus, working only for themselves.

Any company can improve their outcomes by using superior methods. But, unfortunately, even the best companies (or people) do not do everything well. Remember, success is a relative, not an absolute measure.

Many best practices in construction contracting have yet to be convincingly demonstrated as such, even though someone is convinced. Realistically, every new approach introduced into a firm's operation will have ripple effects on other areas due to the complicated nature of our industry. Therefore, evidence of viability should be demanded by construction contractors contemplating the adoption of new practices.

If a firm is focused on improvement, the first opportunity is improving compliance with existing processes. Given the data above, dramatic increases can be realised. We believe that construction contracting can be straightforward or complex, depending on one's perspective and approach.

Two Kinds of Processes

The processes must be formulated on two different fronts: *project and organisation*. They do not conflict with each other, but to separate them is to clarify them.

Project Processes. These methods involve all the tasks that must be completed to comply with all project demands. They can be categorised as follows:

a   Planning.
b   Forecasting.
c   Scheduling.
d   Communicating.
e   Executing.
f   Measuring.
g   Adjusting.

Organisational Processes. These approaches involve all the tasks that must be completed to keep the organisation profitable. They might be categorised as follows:

a   Strategic.
b   Business.
c   Work Acquisition.
d   Human Resources.
e   Financial.
f   Technology.

Technology is listed last purposely; the processes of a – d need to be clarified and implemented before technology supports them. Otherwise, an inefficient misalignment will occur. Practitioners label this "Paving the Cow Path". The cows do not move any faster and are not any more valuable but are cleaner in the process.

Project and Organisational Processes flow one way – from the project to the firm. Projects are the product. Without profitable projects, there is no organisation profit. On the other hand, gross margin ensures office overhead is paid. In addition, net profit allows owners to reinvest in the firm for its further enhancement.

If one is to execute those processes well, then compliance is critical. Compliance with to process is critical in the Lean Approach. Lean advocates its measurement:

• The task(s) should be done by the person(s) assigned to them.
• That the person is logically in the position to perform the task.
• The task(s) was/were done by the person(s) assigned them.
• That the task(s) were done.

Measuring compliance or the comparison of "should" and "did" allow for an in-depth investigation, then asking "why" induces further problem-solving. For example, one study noted, "Almost 80% of the misses were to lack of materials and drawings". This kind of insight allows management to focus on specific areas and implement countermeasures, thereby efficiently getting to the root causes.

Interestingly, the principle of compliance to process is not articulated explicitly in Deming's 14 points of Total Quality Management.

The Power of Adherence to Practices

Consultants and booksellers tout the latest management fads as "breakthroughs". It may be analogous to a new drug. The promise is there but does it have the power to cure? Only after hundreds of case trials can that be answered. But, of course, we are talking about the human body – a complex and sensitive entity. Given the newness of any drug, unproven as it is, would you give it to a loved one? The answer may be "yes" followed by "only in the direst circumstances".

So, it is also that new, unproven practices or techniques are most attractive to the most desperate. But, as complex as a construction contracting organisation, unproven methods have ripple effects throughout recently won projects and those being built. Certainly, executing known strategies that have known consequences is where most senior executives would start.

So, we have previously established the logic via a case study of adherence to practices. However, we have not quantified it. So, let us share its power.

Author Stevens researched this as part of their dissertation. It was an attempt to quantify the power of adherence to company practices.

The adherence data was derived by surveying each company's personnel for their perception of on-time and quality execution of company practices. We used two Likert scales ranging from 1 to 7 (7 being excellent) to measure the importance of a method and the adherence to timely execution.

As you view below, only practices rated high in importance (5, 6 & 7) were correlated. The percentage ranged from slightly less than 50% to more than 90% on time. (Figure 2.10).

In our chart, Construction Organisational Productivity is expressed by the ratio of Overhead/Direct Costs (OH/DC). This ratio displayed on the Y-axis is compared to a company peer average in banking data. That percentage shows a relative standing.

Overhead and direct cost is the sum of all costs of a construction firm. These are the two most significant numbers and do not change much unless they are truly exceptional events. We will not list all the granular expense items that are in each here. Suffice to say that safety or the lack thereof will affect this ratio. So, this is more than people's productivity in the traditional sense.

OH/DC is a straightforward way to look at efficiency. Jim Collins, Author of Good to Great, states that the great companies use a single metric to measure performance in their research. We use this ratio in the same spirit.

There are other productivity measures. We have considered many. They involve many outputs such as net profit, equity, person-hours, gross profit and the like. However, each has weaknesses when compared to Overhead to Direct Cost. Higher than average productivity is a combination of efforts between a contractor's home office and the field staff managing the company projects.

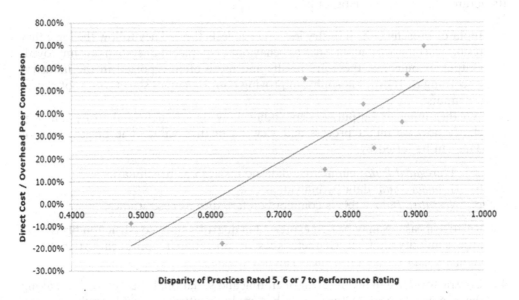

Figure 2.10 Regression Line Showing the Correlation between Adherence and Efficiency.

The chart shows the power of adherence to company practices: there is an approximately two per cent effect on efficiency for every one per cent of completion efficacy. That is true in both directions. Less timely execution of company practices will cause a dramatic efficiency decrease.

As you may have surmised, this ratio makes sense in a highly chaotic industry such as construction contracting. With all the uncontrollable factors and interdependencies of actions, unfinished tasks will delay multiple tasks from being done. This appears to cause rushing at the end of projects, contributing to costly mistakes and reputational harm. On the other hand, according to the data, high adherence has the opposite effect.

Exploring Key Performance Indicators

It is important to note that Key Performance Indicators (KPIs) only emerge from available data and a discussion of any metric's value. We can only utilise the available data differently, and our measurements must be tailored to the Built Environment. It may take extensive discussions to determine what information and metrics can furnish valuable insights to improve local industry performance.

It is essential to note the value of a KPI review. Performance measurements have not kept pace with the construction industry as projects have become more complex (Yang et al. 2010).

KPIs can be absolute or relative measures.

a   Absolute measures are very often numbers, including percentages ranging from positive to negative.
b   Relative measures are compared to a standard, benchmark, or range that gives it context.

Its nature can categorise data into:

1   Input or proactive: the industry less often collects these. They follow the logic that results will improve if inputs such as behaviours, equipment or systems are improved.
2   Output or outcome: these are results and more often reported in the industry. According to TQM philosophy, improving poor outcomes starts with root causes (or inputs).
3   Whether input or output, understanding how data is delivered to us can help clarify how we might process and view its relative value. The presentation of data can be sorted by:

   a   *Aggregated Data:* cumulative data points such as benchmarks.
   b   *Granular Data*: data by individual case.
   c   *Narrative Data* – this is commonly referred to as case studies.

   Relative measures may be best expressed in quartiles since construction firms have little similarity in their characteristics or operational approach. Attempts to compare exposes the problem of intersectionality.

4   Contractors have many different operating and situational characteristics. Taking the top ten critical factors, each has multiple attributes and results in over

1 million variations. This is critical since the industry employs approximately 385,000 Built Environment organisations.

Projects have many dissimilarities also. We call it *"intersectionality"*. Numerous factors, such as location, project team composition, schedule timeline and technical complexity, make each project unique.

5   Absolute measures might be presented against a range of values, including least, most and median. This will give context to the measure's meaning.

6   For the built environment, the median is a better measure due to the industry's uncontrollable circumstances. Median is more representative than average in construction due to the widely varying factors such as job site conditions, design extremes, economic climate, weather and contract type. Additionally, input errors and "pencil whipping" do sometimes occur.

Our General Recommendation

There are many uncontrollable factors when constructing. The industry's dynamics are unforgiving. Once an adverse event occurs, its effects ripple to multiple downstream tasks. So, a careful approach to KPIs is to measure inputs such as behaviours like planning quality, resource forecasting and schedule prediction. There are many more. These leading indicators distil proactivity intensity will add further safeguards against adverse events.

We recommend adding proactive measures to existing output measures.

The three mentioned – planning quality, resource forecasting and schedule prediction – will be explained further to define further and illustrate this thinking's value.

Planning quality: detailed and holistic documentation of aligning installation of construction work signifies a less risky approach: perfected BIM models, realistic CPM schedule, Rummler Brache project maps and Project Executive Letters of Instruction prompt deeper thinking than average to minimise oversights and blind spots in planning.

Resource forecasting: Main Contractors and subcontractors do not build one project at a time. Therefore, there is a value to forecasting resource needs across all their projects that may show resource shortage against demand six weeks in the future. If discovered six weeks in advance, it is a challenge. However, the same problem found today is a crisis. Common resources forecasted – supply versus demand each week for six weeks into the future – are labour and craftsperson, management personnel and shared equipment.

Schedule prediction: the less late task completion between a baseline, rescheduled and actual Schedule (Linear or CPM), the better planning and control a construction organisation manifests, the better the quality of those critical management functions.

Most construction firms have a strong interest in productivity measures. However, it is important to remember; productivity measurement has three critical features: 1) accurate recording by the site; 2) physical completion/labour-hours expended comparison; and 3) projection of final outcome. Keep these in mind as you create or refine your productivity reporting processes.

Certainly, any construction company that deals with active or bureaucratic clients will need some kind of system, including trusted measures. Contractors do not need a failsafe method, but they do need an organised one. Each contractor must determine how best to measure (reports) and store these reports (documentation). Electronic filing systems serve this purpose well. (Table 2.5–2.8).

Table 2.5 Examples of Built Environment Data and Possible Sources

Data	Possible Source	Comment
Safety	Government & Insurance Firms	Both have an interest in more wide-ranging and in-depth data collection and analysis. Research should result in industry insights to help all stakeholders
Employment	Australia Bureau of Statistics	Australian based data such as Demographic, Geographic, Employment type, etc
Productivity	Australia Bureau of Statistics	The ABS Multifactor Productivity Index (longitudinally) is a readily available dataset
Quality	Private Testing OrganisationsBuilt Environment Firms	Non-compliant installations and products cause rework which is a significant contributor to unproductive time
Schedule	Construction Service Buyers	From individual projects – Baseline versus Actual (schedule deviation)
Planning	Built Environment Firms	Types and quality of planning documents, including contingency plans
Contract type	Government, Industry and Legal Organisations	Alliance versus Lump Sum versus Negotiation versus IPD etc
Waste	Government or Industry Organisations, including Waste Management companies	Waste tracking by source, type, amount and disposal disposition appears to be paramount
Relational	Project Specific	People engaging People such as meetings assessments, site-office conflict surveys and collaborative teams' measures
Financial	Robert Morris Associates	For comparison only – some measures are focused on risk assessment derived from financial ratios
Financial	Australian Stock Exchange	Typically, macro numbers for a publicly held company per stockholder's report

The Single Measurement Challenge

It is critical to have a preliminary or primary metric with many KPIs to choose from. Indeed, construction executives are busy people. Additionally, all projects, clients, teams and partners are different. We don't recommend that a contractor has tailored metrics for each one. That would take some effort and would increase confusion. Sometimes, multiple measures might start an argument from a manager about their performance. The manager could emphasise the positive KPIs and deride the others.

As we stated, some measurements are relative and some can be absolute. We would note that there is little data in Australia that would give confidence to an absolute metric. So, we are practically left with a relative metric in most cases. But, of course, bottom-line profitability is an absolute measure.

Table 2.6 Possible Categorisation of Data

Nature of the Data	Example	Comment
Proactive	Practice adhered to or behaviour exhibited	Best practices executed are precursors for improved results by definition
Reactive	Outcome	Output measures per time interval specified
Absolute	Number	Typically, it can range from a significant negative to a large positive number, such as the project's cost or schedule deviation. For example, rework and defects would start from zero but can number in the hundreds
Relative	% Adherence against standard trend line may be considered a relative measure since it uses previous periods for comparison	Average Contractor Safety can be measured as 1.0 and the participating contractor safety experience can be measured higher or lower than 1. Market share is best performed on a trend since data collection and the measure may be flawed

Table 2.7 Input Data Examples

Example	Comments
Practice Adherence	Practice examples include Planning, Communicating, Resource Forecasting, Executing, Measuring & Adjusting subtasks
Practice Listing	Few organisations have a listing and articulation of standardised practices
Practice Measurement	Monitoring and measuring timing and adherence
Attendance Data	All stakeholder's management attendance percentages and workers promised/actual to the site per day by subcontractors
Conflict Activity	Worker complaints and legal meditation actions in a project
Safety Behaviour	Notifications of unsafe conditions and behaviours
Cycle Time	Submittal approval time or Crane activity durations

Table 2.8 Outcome Data Examples

Measured Outcome	Example
Project Quality	List of defects and their percentage to overall project schedule tasks
Project Schedule Outcome	Baseline versus Actual, including new items such as variations and rescheduled baseline items
Project Safety Outcome	Number and Severity versus industry averages per hours worked
Project Cost	Estimate versus actual, including early predictions of the final cost
Project Resource Allocation	The capture of weekly management, labour, equipment and cash allocation decisions
Organisational Adherence to Practices	An organisation's endorsed practices are measured for adherence
Project Waste	Weight and types of waste and disposal
Stakeholder Perception	Comparing practitioners' typical project experience to a specific project experience

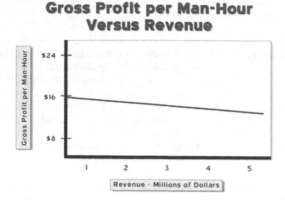

Figure 2.11 Field Hours versus the Gross Profit They Produce.

As we have stated, the field operations are where there is 100% of turnover, 90% of cost and a majority of risk emanating from. So as a further and refined measurement method, we have shared two field oriented productivity metrics. See Figures 2.11 and 2.12.

Figures 2.11 and 2.12 could be accurate measures indicating relative success or not. The first is Gross Profit per Person Hour (GPPH). It shows the profit-producing efficiency of a project, division or company has. Field hourly labour is the great wild card in construction. If you have many efficacious practices, the trend line can be steady or increasing, all other external factors notwithstanding. This is an example and the level of profitability is for example, only. As you see, as the volume of work rises, the gross dollars may decrease. As we stated earlier, this is quite normal; many contactors target gross dollar goals rather than profit percentage in their tendering calculations.

The second single measure we would like to share is slightly different and for a reason. Main contractors many times employ salary staff exclusively. This salary staff

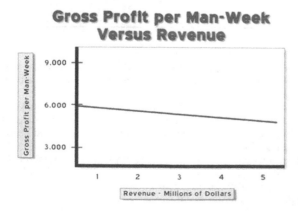

Figure 2.12 Person-Weeks and the Gross Profit They Produce.

Table 2.9 Matrix for Field Operations Comparison

Year	XXX7	XXX6	XXX5
Labour Person-Hours	256,400	220,560	204,200
Project Managers	10	8	6
Field Supervisors	13	10	9
Total Gross Profit	$7,121,696	$5,955,866	$4,503,877
Gross Profit per Person Hour	$27.77	$27.00	$22.06
Gross Profit per Person Week	$5,955	$6,363	$5,136

is not paid hourly, nor should they be tracked hourly. So, a week is a divisor and a gross profit per person week (GPPW) is the measure. Again, as turnover increases, this ratio will decrease. Speciality contractors that employ hourly labour may use both as measures since they have both kinds of payroll arrangements.

As a further example, how this might be reported. See Table 2.9.

If you calculate, you will see that the increase in gross profit outpaces the growth in people count. The company has an excellent trajectory in its business when GPPH and GPPW. Year XXX5 is $22.06 and $5,136. In comparison, Year XXX7 is $27.77 and $5,955, respectively.

The takeaway is the importance of determining the "one metric" as a starting point for the first indicator of a problem or opportunity. This straightforward approach efficiently manages several situations, including vendor performance and client collaboration. Then, if needed, a closer inspection will reveal the total picture and you will be ready to act with confidence.

3 Work Acquisition

In construction, efficiency is a second priority to work acquisition. Building and managing the project pipeline is the prime concern for the construction managing directors. This is a reality in an industry that has historically experienced rapidly decreasing project opportunities. Many contractors know examples where competitors went bankrupt while becoming more efficient (Green 2011). Work acquisition keeps valuable employees working and producing turnover, covering their costs. Construction's single-digit profit percentages do not allow for stockpiling under-utilised or unproductive employees.

One critical precept of work acquisition in construction is not chasing poor relationships. In all businesses, people will treat you with differing levels of respect and care. Find out early where you stand. If no connection can be built, move on to the next potential client. Some clients may see you as "just another contractor". The good news is that you only must find a handful of clients who feel you deliver value to have a profitable construction business. Do not chase potential customers who treat you as "just another construction firm". As you know, these clients tend to be unprofitable.

The number of qualified leads you follow up on is the reason you will land a good project. Keep conversations going with decision-makers and have budgets, needs, wants and with those who treat you with respect. At some future date, they will have a deadline to meet for constructing a project. Keep yourself at the client's table. Talking does not cost you a dime.

Estimators are becoming more of the project team. This was in contrast to years ago when the take-off and pricing were done in a relative vacuum and then presented to a senior manager, usually the Chief Estimator, for review and discussion. With the advent of increased competition and the design construct concept, more companies are making the estimating team part of the ongoing business process. Estimators are asked to help sell the projects and interact with the field to verify effective construction techniques at the tender time.

Contractors Have to Win Four Times to Win a Project

Construction is one of the few industries that require demanding qualifications of businesses to operate. Since high risk is part of the fabric, regulators and construction service buyers make contractors provide extensive information to prove their capability and mettle. The growth of risk management as a discipline appears to be a cause of even more scrutiny of contracting firms.

DOI: 10.1201/9781003290643-3

Our view is that construction company executives must win four times to win a project. To qualify their firm to propose to build shelter or infrastructure:

1 They must pass state or national licensing requirements and, many times, financial examination. This requires:

   a Site experience
   b Superior professional behaviour
   c Knowledge, both business and craft/technical
   d Capital
   e Character
   f Capacity

2 Municipal licensing and business requirements are more demanding than ever before. Site experience, testing and financial requirements must be satisfied before a contractor can work in many locales. Several metropolitan areas of the country demand more qualifications than the states in which they reside. As a side note, "outsiders" to any location may be tested by public and private interests to protect the "locals".

3 They must qualify for a "shortlist" with the client (construction services buyer or main contractor) by interviewing in-person and providing information concerning:

   a Past projects.
   b Staff experience and qualifications.
   c Financial statements.
   d Insurance coverage.
   e Financial capacity outside of the organisation.
   f Other important factors.

4 Then the process proceeds to a specific project focus. Contractors must provide value that matches the client's needs and wants. Of course, some clients only desire price. Others prefer service and quality at a competitive price, especially if they occupy the proposed residential or commercial building. Meeting specific needs and wants requires:

   a Capability to perform a particular project; this could include employing experienced craft people, site supervisors and project managers
   b Ability to identify client desires while managing critical success factors about the project
   c Experience in the upcoming job type
   d Past work with similar clients
   e Price with terms and conditions

In the area of giving the client-specific value, it is typically not a question of price. Instead, when superb contractors compete against their peers, fair profits, terms and conditions are proposed.

So, what are takeaway ideas here? First, contractors who work hard to make projects come in on time and budget create a positive future for themselves. This is not a corporate business; it is an owner-operator one. Construction professionals make their luck. The owners of above-average construction firms do not live in their offices; they visit jobs and are on-site decision-makers. In essence, most time is spent outside running the business, keeping people productive and projects profitable.

Younger contractors should not be impatient about winning a negotiated project. They must win four times to do so. Strangely, an inexperienced construction company might perform better on the fourth challenge if the client is focused solely on price. However, without qualifying first, a contractor can never sit at the negotiating table with the client.

As a help to direct the young constructor, targeting projects requiring financial security is a savvy move. Many companies' total value is not significant and erects a barrier. In other words, this is a niche. In general, a significant percentage of contractors are somewhat undisciplined in their financial management. On the other hand, contractors with solid banking relationships are typically very well organised and disciplined in their finances, among other things. Ask your bank; this is an underrated niche. It is perhaps the most robust niche.

Is Successful Work Acquisition Mostly Art or Science?

Work acquisition is critical for all contractors. Consistently winning work means building work and keeping personnel employed. If a contractor cannot acquire work, they may be just a pricing service to disinterested clients. Pricing work profitably means knowing your costs. Earning and collecting margin on jobs means a contracting business is profitable. Placing accurate and detailed proposals in the market gives a construction firm the best chance for continued success.

This article concerns the process of winning work and one approach to this challenge. The successful methodology is based on numbers and algorithms.

Cost estimating in an inconsistent and unreadable fashion causes many problems which ripple through to the CPM schedule, including lack of information for site staff and tracking ability for the company owner once the project is won. Additionally, construction contracting is a one-way street. You only win when your price is lowest or competitive and not highest. There is no balancing effect on the total number of tenders. All these factors threaten the turnover dollar with lower margins and slower turns.

Below are steps for determining accurate cost and appropriate profit margin on a tender.

Market Selection

Narrowing the type of projects to tender is the first step. No company is a master at all custom projects in a complicated business. Grading each factor such as project type, potential client and possible location allows a managing director and other company leaders to quantify the opportunities' quality. For a clear understanding of the best location-client type-project type, many factors should be assessed for each potential client. This should include questions such as difficulty to qualify, payment lag, competition strength and focus on quality versus price. There may be more than a dozen. This system is numbers-based, determined by executives who assess the alignment of each opportunity with your company's skillset.

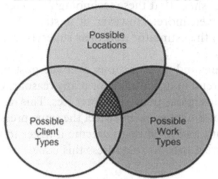

Effective Market Focus

Figure 3.1 Market selection is composed of three primary components.

Construction contracting organisations that select the most advantageous geography, client types and project categories are more often successful in winning profitable work. See Figure 3.1.

Project Qualification

As an initial step, the number of qualified leads a contractor earns correlates to proposal success. If the firm has many skill-aligned projects (ones they can construct well) to price, they will win something profitable. On the other hand, having just one and you cannot walk away if the price is the client's only selection criteria, or the terms and conditions are onerous.

Go/No Go

When Requests for Tenders (RFT) are received, many firms then use a Go/No Go process to prioritise the order of projects to propose. It is a graded approach with specific factors rated. The amount of tendering resources is limited and should only be invested in good opportunities. Some of these RFTs might be not answered due to one or several high-risk factors a project possesses.

Quantity of Work

Quantity take-off is one of the most tedious and risky steps of contracting. It demands that the number of work units be derived from pictures, written words and observations. It must be accurate. Zero times any number is zero. Using an overly complete checklist (constantly updated) and having a triangulation method minimise counting errors.

As we know, words are inaccurate. A picture, or a set of plans, is worth a thousand words. Still, construction plans are becoming less detailed; so starting the RFI process in the take-off stage and pursuing it until submission day is the only effective tactic for eliminating inaccurate assumptions.

Technology has improved accuracy, consistency and speed significantly. In addition, tools such as Digitisers, Estimating Software, Assemblies and BIM cut hours out of the quantity take-off process. Some studies show that these technologies reduce the time spent to half; others' research report even more. However, it is still a human endeavour. To ensure excellent performance, the estimator and their supervisor must employ a thoughtful and disciplined process.

Our observation is that the quality of plans and specifications over the past three decades has diminished. Unfortunately, this trend may continue for many reasons. One is that design groups are forced to produce plans in less time with lower fees. This results in less time spent in the creation and review of documents. They feel the same pressure contractors put on a lower price for their work and promise to execute it faster to win projects. Construction professionals must have a process to address this issue.

Cost of Work

Without accurate labour and equipment productivity numbers, a company will propose a price too high or low. Either situation creates undesirable consequences. The contractor will get work at a loss or not win any work due to higher than market pricing. Reliable and predictable cost data can be derived from several sources, including:

* Site experience,
* Observation of crews, or
* Company cost records.

Your previous projects' cost performance data are the best. After a couple of dozen job cost entries, there is a creditable cost history to start your thinking. Certainly, dissimilar projects produce a range of labour and equipment productivity outputs. In our experience, the median number in that range should be used. The high and low quartile of the data contains errors such as incorrect coding and extreme job site conditions.

Indeed, your cost figures should be unique to you. The use of cost services such as Rawlinson or Cordell is a last resort. These services can be valuable and should be used in certain circumstances. However, day in and day out, your cost structure cannot be public information.

Unit pricing must be for what you do work. That is your actual cost or production rate. It is hard to justify "what our best people should do it for" or "what we did it for in the past years". The margins we work with do not allow for that.

General and Administrative/Overhead Cost Application

This expense can add 10% or more to a tender where the competitors' price proposals are within 1% of yours. As we know, the labour content of a project directly affects general and administrative/overhead costs. In other words, labour and equipment cause companies to hire payroll clerks, safety professionals, shop managers and operations leaders. With each is associated office space, vehicle, technology and miscellaneous costs (Figure 3.2).

Appropriately allocating this overhead is critical. A company's cost must be recovered; otherwise, it may pay for finishing a project.

Budget for 20xx Worksheet

XYZ Construction

	20xx
A. Total Material & Subcontracts Budgeted for Fiscal Year - 20xx	
Direct Material	$2,030,000
Direct Subcontracts	$280,800
Total	*$2,310,800*
B. Total Labor, Equipment & GC's Budgeted for Fiscal Year - 20xx	
Direct Labor	$2,405,000
Direct Equipment	$120,000
Other Directs	$210,000
Total	*$2,735,000*
C. Company Overhead to Recover Budgeted for Fiscal Year - 2004	*$676,000*

Figure 3.2 Input Screen for Total Projected Organisation Costs.

Proper overhead allocation depends on several factors. The most important is an accurate yearly company budget. Besides the overall numbers, it must delineate the labour & equipment/material & subcontract amounts of the new year's projects. An accurate direct cost and overhead budget lead to better results (Figure 3.3).

Only a handful of formulas capture the realities of overhead recovery for contractors. The Dual-Rate Cost Application Methodology, an algorithm formulated from construction contractor banking data, is the most accurate. In our single-digit-net-profit-before-tax business, we do not have room for errors or guesses. Said differently, the difference between loss and gain is razor-thin, while the client almost always accepts under recovered cost proposals. Therefore, top-quartile contractors use a method that accurately reflects the management overhead cost of directing various construction work components.

Job Sizing

Once the office cost of supporting and managing the project is determined, a job sizing algorithm is used to determine the efficiency of the overhead. Large jobs' overhead can be lessened. Smaller jobs increased. Interestingly, the smaller jobs are disproportionately inefficient than larger jobs as determined by banking data.

At this point, a contractor should know their breakeven cost. If they know the cost and checking counts, they will be confident of this estimate. From here, an artful process starts. First, the client negotiates for a better price, among other things. Next, the contractor weighs the proposed construction contract to assess the speed required, the margin needed and the scope demanded. As you know, furnishing all three (higher speed, lower margin and extended scope) leads to bankruptcy. Providing two of the three can make for a reasonable agreement.

Profit Margin

What is a reasonable margin on a project? What factors increase or decrease it? There are several factors to consider: project risk, backlog, return on the investment

DUAL OVERHEAD COST APPLICATION CALCULATOR

XYZ Construction

A. Total Material & Subcontracts Budgeted for Fiscal Year - 20xx	**$2,310,800**
B. Total Labor, Equipment & GC's Budgeted for Fiscal Year - 20xx	**2,735,000**
C. Company Overhead to Recover Budgeted for Fiscal Year - 20xx	**676,000**
D. M&S to L&E Ratio	0.84
E. "X" Factor	2.36

F. Rate Calculation Formula
   For Materials & Subcontractors(percent)

$$\frac{\text{Overhead} \times 100}{\text{"X"} \times \text{L\&E} + \text{(M\&S)}} =$$

Overhead on M&S	7.72%	This rate is set for the budget year Put into estimating equation Budget must be an accurate prediction

G. Rate Calculation Formula
   For Labor and Equipment (percent)

$$\frac{\text{"X"} \times \text{Overhead} \times 100}{\text{"X"} \times \text{L\&E} + \text{(M\&S)}} =$$

Overhead Rate on L&E	18.20%	This rate is set for the budget year Put into estimating equation Budget must be an accurate prediction

H. Proof of Recovery

O/H Rate x (M&S)	178,333
O/H Rate x (L&E)	497,667
Total Overhead Recovered	676,000
Overhead to Recover	676,000

Figure 3.3 Example Project Dual Rate Overhead Cost Calculation.

Job Size Overhead Cost Calculation

Average Job Size (Direct Cost)	$122,020	Normal OH to Recover	$21,500
This Job Size (Direct Cost)	$324,420		
Direct Cost-This Job (Ji/Ja) —————————— Average Direct Cost	2.66	This Job OH to Recover	$19,436
Overhead Multiplier (Ctrl-C to Compute)	0.904		

Figure 3.4 Example of Project Overhead Cost Sizing.

required, competition and client coordination ability. Indeed, a return-on-investment calculation should double-check the final margin (Please note markup is not the same as margin).

With each factor, we should be increasing or decreasing our price. Improper pricing makes us miss opportunities or win jobs we regret winning.

Contractors who use modelling give themselves the best opportunity to win consistently. Every contractor would be wise to consider generating a data-driven model to drive future tendering decisions.

The most effective models researched and implemented are the ones including a history of competitor's tenders. As in sports as well as business, all organisations and their employees have tendencies. You, with the use of your model, quietly quantify those tendencies. As a result, you will win more tenders with "less money left on the table" and your competitors might never know what you are doing.

In summary, construction contracting starts at the project qualification stage – also, this is where most of the risk begins. Without a superior estimating process, the rest of the business is more complicated than it should be. The site is where most of the risk, 100% of the turnover and approximately 85% of construction firms' cost. The estimate should help the project staff clearly understand the job's costs, technical details and material needs, among other factors. In other words, estimators should be sensitive to the needs and wants of the site staff.

The work acquisition process outlined is a numbers-based system which makes it mostly science. It starts with a correct count and continues using your unique costs. From there, the science continues from general conditions calculations to overhead costing. Next is bid modelling and lastly, the cut/add sheet filled in the hour before the tender or proposal is due. This final step is where artful processes begin. Hunches and intuition are factors along with last-minute price reductions from project partners. Final contract negotiations continue this intuitive process.

Winning projects means building projects and earning profits. If you cannot win work, your firm may be providing a pricing benchmark for competitors. Indeed, this quasi-scientific process outlined here is a start. It is complicated and the client influences part of it. There is a long process map when put to paper. All in all, the more logic-based and hardwired, the faster and less problematic it is. Making unnecessary subjective decisions in each step only clouds the final price determination with terms and conditions.

Examining Work Acquisition Performance With a Single Metric

Using numbers to determine the quality of an outcome is not new in the business. Profit, revenue, expenses, asset values and the like are all numbers and critical ones. The ultimate financial number is net worth. It is essential to realise that numbers are facts and not emotions, opinions or thoughts. Like net worth, facts are definitive and verifiable. Using trusted numbers to discuss outcomes, opportunities or actions is a valuable quantitative method. This article discusses the use of trended market share and gross profit as a macro measure of company performance.

Business development and work acquisition professionals pay close attention to market share and gross profit percentages. They know that these two are rough indicators of their performance. Therefore, pricing work to maximise profit and then win the proposed work is an overall job description. To define and clarify terms:

- **Gross profit** is what remains from turnover after direct or site costs are deducted. Some also know it as contribution margin. The resulting gross profit is logical to track since contract price proposing and direct cost prediction are the two largest financial controllables.
- **Market share** is a calculation of either value or number of projects won divided by the total available. Using value or number depends on the situation. Market share can be thought of as a baseball/softball batting average, i.e., the number of hits divided by the number of at-bats.

This gross profit percentage is tracked on a trend line and market share (whether number or value of projects). The comparison of the two shows an essential result of the contractors work acquisition efforts: the ability to gain work and profit margin. In addition, a trailing three-year trend gives insights into the efficacy of your people and processes.

What is the interval timing? We have seen monthly and quarterly trend lines. Therefore, if the calculations are on a monthly trend line, there would be 36 data points at any one time. (three years multiplied by 12 months). On the other hand, if quarterly calculations are more realistic, 12 data points would be trended (three years multiplied by four quarters).

Taken together, trended market share and profits indicate whether we should be satisfied or concerned with our work acquisition performance.

For example, in poor economic times, your batting average or market share should stay the same as your historical average. Your profits may go down and the number of opportunities may be fewer, but you should be able to close the same percentage of contracts. Even when the market gets tight, the company's proposing effectiveness should not change.

Comparing profit percentage to market share tells a story and may prompt action. Without comparing the two, a problem can be brewing and may fester until it becomes a crisis. Use the two metrics together to stay ahead of potential work acquisition problems.

Said differently, we believe that comparing the contractor's market share with gross profit percentage helps to indicate successes, failures and concerns. Some typical market share scenarios are shown below. Figure 3.5 illustrates what happens when the contractor attempts to "buy" market share by cutting profit margins. This action results in working harder for less profit.

Are your clients demanding more services but not paying you for them? Answering questions such as these and taking corrective action will help you pull out of this concerning situation. Some construction service buyers are overly aggressive about cost and may use threats (of future work, legal action) to decrease the margin you can charge them.

If you see your market share stagnating while your profitability is losing ground (see Figure 3.6). It would help if you investigated the reasons. For example, is the economy poor? Thus, you have had to cut your margins?

Some believe a rising market share spells success. Others say that market share should stay level while profitability is the key measure. As you know, opinions abound in our industry. However, we believe that if profitability is non-existent, other measures do not mean much.

The third scenario shows a falling market share with a corresponding reduction in profitability (see Figure 3.4). This represents the most stressful business scenario.

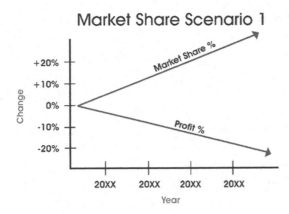

Figure 3.5 Market Share Rising but Reduced Profit Margins.

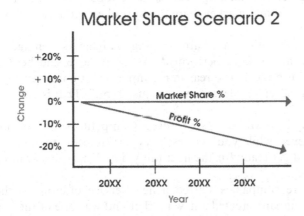

Figure 3.6 Market Share Stable but Profits Down.

You are winning less often at a lower profit percentage. The reasons for this include poor craft skill, bad project execution, negative client perception and higher competition. The situation shown in Figure 3.7 is reminiscent of the 2008–2009 Great Recession.

Your batting average is declining and you are being paid less. Essentially, you are doing more work in estimating, winning less and getting paid less for it. To reverse this trend, you must determine and attack the reasons. For example, pay better attention to quality, improve project execution or differentiate your firm in unique ways.

For specific construction sectors such as low voltage, test and balance, service, drywall, acoustical ceiling and the like, we have found that it does make sense for contractors to have a market share percentage, but on a three-year trend basis. The last data point in any trend measurement is today's market share. This provides currency as we use the same measure of the market against our sales. A substantial change in your market share signals something important about your business.

Figure 3.7 Falling Market Share and Profitability.

Trending the market share makes sense, especially in a changing economy such as a boom or recession; it keeps the decision-making process more focused on the facts. Whether the economy is rising or falling, market share measures keep you away from a knee-jerk reaction.

Figure 3.8 illustrates a golden situation. You are winning a higher percentage of your bids and earning more profit for work performed. As many have said before, knowing why you are successful is critical. The reason is simple: so, you can repeat it in the future. The answer should never be "because we are just good". That is a recipe for future problems.

If this happens for you, enjoy it. It will not last forever. Competitors are not un-observant or complacent. However, while your success is occurring, save the profits, reinvest some in the business and get ready for the next rainy day. Rainy days can be harsh in construction contracting.

We have seen situations where companies started a new service offering to the market. Each was novel. Each company created a new market and was one of the few

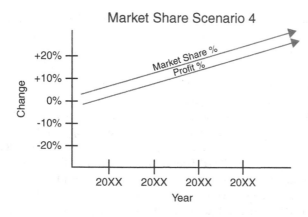

Figure 3.8 Rising Market Share and Rising Profitability.

(or the only one) to provide the service. As a result, its market share stayed over 50% and the market dollar volume increased. It is an enviable place to be.

Market share may not be helpful for contractors who may construct large projects such as Multi-National Specialty Contracting Firms or Construction Management Corporations. The problem is that there are such wide swings in volume and thus market share as each project is won or lost is that the trend line will dramatically move up and down. Also, when these kinds of firms are full of work, they back off their aggressiveness. As a result, their trends are difficult to interpret.

There are several iterations of the market share and gross profit interaction. Table 3.1 shows this relationship on a three-year trend basis. This is our summary.

The ratings are explained as follows. These are our labels and definitions for particular construction contracting business situations.

Golden: This is the best of all possible worlds. The business is gaining ground on the competition as indicated by its market share and being paid more for it.

Business Savvy: The company is keeping market share level while finding more ways to profit from operations.

Harvest: Contractor may be taking the opportunity to price higher, lose market share, but gain profits.

Future Plan: The company is keeping the profit percentage level but also winning more work. At a future date, this can be "harvested" by raising the profit percentage, winning fewer projects, and making more money per person-hour.

Day-In-Day-Out: The company is not gaining or losing in both profit and market share percentage in either area. This is a long-term approach; they simply wish to build wealth and value incrementally as the owners reinvest some of the profits and distribute the remaining to shareholders.

Red Flag: Profit percentage is steady, but market share percentage is falling. You need to ask whether you are losing your aggressiveness. Are businesses or customers leaving you? Would it be a good time to make some personal calls?

Buying the Future: You are gaining market share percentage but making less profit in doing so. Is this your purpose? Are you running the wheels off the trucks? Are our people getting tired? Will you at some future date reverse this and make more money working less?

Market Reversal: Your market share percentage is level. You are hitting your batting average. However, your gross profit percentage trending down since you have been forced to bid at lower profit levels in recent months. The economy may be weakening, or a competitor may be driving down prices. This is a frustrating scenario that needs further investigation, thought and discussion.

Depressed Leader: Losing market share and profits. The leader may believe that the economy is poor and is acting as if that belief is true. If a new, unsophisticated

Table 3.1 Three-Year Trend on Profit and Market Share

	Market Share	*Market Share Steady*	*Market Share Falling*
Profitability Rising	Golden	Business Savvy	Harvest
Profitability Steady	Future Plan	Day-In-Day-Out	Red Flag
Profitability Falling	Buying the Future	Market Reversal	Depressed

competitor is in the market, the manager may compete against this thoughtless contractor. You need to do a brutal analysis of this situation.

Your business development person may not like market share and gross profit measurement. This is a common reaction from employees as these places added pressure on performance. However, we are all measured in our professional lives; business development and work acquisition people are no different. Our observation is that what will appear is a dramatic difference among individuals in this area. The best business development and work acquisition professionals take this measurement as a challenge. The same can be said for any worthy supervisor or project manager. The better leaders rise to meet a challenge.

In all this, this creates a system of behaviours, internal infrastructures and measures to better run their businesses. In other words, a culture representing the proper way to work and thus to be rewarded.

This is part of a new management model for construction. We can no longer rely only on our gut feelings and experience. The construction industry is changing, complicated and riskier.

One of a dozen processes moves a construction business to its "sweet spot" of minimising risk and maximising reward. This quantitative method helps give clarity to the challenge of managing a contracting firm. Contrast using this with the trial-and-error process used by the older generation over their career. A practice- and results-based method is faster in determining a beneficial course of action than an experience-based one. Simply put, this helps lessen the waiting and pain of your Grandparent's practice.

Tendering Smart: Market-Based Pricing Strategy

Contractors must do a dizzying variety of tasks well. These tasks span the three components of the project work cycle: Acquiring Work, Constructing Work and Keeping Track while attending to the wants and needs of the staff who execute them. In addition, this section provides insight into market-driven strategies to help contractors decide when to bid, what price to bid and when to pass.

The worst possible outcome for a contractor is winning the bid at an unrealistically low price. Taking on the risk of a construction project for no compensation is hard to justify unless done in extreme circumstances. However, contractors who pay attention to data and the current market climate in their pricing and tendering can be more confident that the bid they submit optimises profit.

Tendering Smarter

From Marvin Gates' work in the 1960's to now smarter and more profitably has evolved from being mostly an art to more of a science. For example, contractors over their careers have been frustrated by "outguessing themselves" on bid day. This resulted in either 1) leaving too much money on the table (out pricing their competitor by a large margin, or 2) narrowly missing winning projects (losing too many bids consistently by small margins). However, that can change with the use of structured thinking, which guide the contractor in taking the best risks. This section provides an example of such structured review that is robust and importantly, teachable to the next generation of leaders and managers.

Why else should contractors who have been getting by on "feel" move to a structured approach to tendering? Because we are a single-digit-profit business and small adjustments make big differences to our return on investment (ROI). Also, it is not only about winning the job: it is about winning the *right* job – the right job for the firm at that particular time, the right job for your aspirations and plans. Another good reason to move to a structured approach is that other highly profitable industries do it.

One highly profitable industry, Insurance, reviews many data to decide whether your characteristics – such as age, smoking and employment status – make you a good gamble for a policy and at what price they should charge you. Casinos are another example. These two consistently average the highest profits over many years over most others.

Insurance companies and gambling casinos cannot predict if you as an individual will win or lose (die early or live a long life). They can, however, use structured statistical approaches to minimise their risks. Over many repetitions, they have worked it out to usually win small or large on a high percentage of clients. Contractors should consider the same.

Data: The Foundation of the Structured Approach

The linchpin of the structured approach is information. To use a statistical process for tender behaviour prediction, which we will use to guide our bid pricing, we need to capture some simple data from each bid. Each bid contains valuable information for construction contractors to capture and use. Table 3.2 shows a hypothetical set of proposals for a single job by contractors.

In Table 3.2, see the following columns:

Competitors – They are your like-minded and capable equals. They approach the business much the same as you do with ethics, qualified crews and a reputation for superior work. They, like you, will be around for a long time.

Bid price – This is available easily if bids are on public work. If not, there should be a friendly owner or contractor who will give you the bid results. If not now, maybe later.

Cost of the work – You will never know your competitor's true cost; however, you know your project cost.

Margin $ – This is the gross amount difference between your cost and their bids.

Margin % – This is the percentage difference between your cost and their bids.

In attempting to create a tender model, we know that we need much more information than a single project to extrapolate trends. You can use many tables such

Table 3.2 Results from a Single Project Tender Opening for Peer Competitors

	Bid Price	*Cost*	*Margin $*	*Margin %*
Simonelli	$150,000	x	$18,433	14.01%
Post	$155,000	x	$23,433	17.81%
Sabah	$152,430	x	$20,863	15.86%
Almeida	$149,678	x	$18,111	13.77%
Gillard	$151,000	x	$19,433	14.77%
Our Company	$154,000	$131,567	$22,433	17.05%

Table 3.3 Aggregated Bid Results of Peer Competitors

Company Name	Median Mark-Up %	Lowest Mark-Up % Above Our Cost	Highest Mark-Up % Above Our Cost
Simonelli	14.13%	13.94%	15.78%
Post	15.21%	14.02%	16.32%
Sabah	14.78%	13.68%	15.99%
Almeida	13.78%	12.11%	14.89%
Gillard	14.99%	13.44%	16.03%

as Table 3.2 to compile another table like Table 3.3 below, with aggregated results for many tenders. We suggest collecting a minimum of 30 projects and these should be relatively recent. After more than 30 bids in which your firm and your competitors have participated, an aggregated bid results data set can be created and used to understand others' pricing behaviour.

What is the history of our competitors? Take a look at the columns below. Each competitor has a:

- Median % of markup. We do not use averages because they can contain outliers that distort the metric. Also, sometimes competitors have bid errors such as missed quantities and math mistakes. These are not qualified bids. The same could be said for quotes when your peer is busy and submits a high number. The median is a better representation of the "typical" bid.
- Least. This is the lowest responsible bid the competitor has submitted.
- Most. This is the highest responsible bid. Good information if you do not want the job.
- Other factors. Additional data should be collected for contractors who genuinely want to work smarter, such as 1) gross profit dollars, 2) designer of record and 3) governmental jurisdiction. Attributes such as these will show tendencies of your competition without them knowing about it. For some trade and speciality contractors, this is an enormous amount of data to analyse. Sometimes, a summer intern starts this data collection process for a construction firm. This data represents a significant opportunity to capture additional profit each year.

In the next section, we describe the general logic of the bidding strategy. In the following section, we use Table 3.3 to clarify the rules of bidding that many high-performing contractors utilise.

The Five Rules of Tendering Strategy

From our research, we have summarised bidding in five rules with which to guide contractors.

First Rule – *You cannot compete against irrational behaviour.* If you do, you must become irrational to win against it and you will suffer business consequences for this action. What you can do is cherry-pick the opportunities until the competitor changes. That is rational.

Second Rule – *Since the most profitable industries use statistics to predict long-term behaviour, we should use it to predict competitors' moves.* People and companies have general behaviour patterns. These are consistent over time. People are never entirely predictable in a single circumstance; however, long term better results against your competitors as these behaviours repeat over hundreds of bids.

Third rule – *Competitors have different values and, thus, different strategies.* However, most people will be consistent in these strategies, even if flawed. This is one of your competitive edges.

Fourth Rule – *The higher the price, the lower the hit rate.* This is a business rule that will reward you. If you can increase the amount you bid accurately; you can increase the amount of profit.

Fifth Rule – *Your competitors who are equal to you will raise and lower prices with many of the same factors you do.* Numerous examples exist, such as the amount of backlog they may have or the complexity of a construction project. In the aggregated data, dozens of tendencies such as these will become apparent.

Together, these five rules form a structured approach likely to yield results that maximise your ROI and minimise your risks. Let us now look again at Table 3.2 and see this structured approach in action.

Application to Tendering

Tomorrow's bid should consider the lowest and highest responsible bids, above our cost. Start with your estimate for the job, then compute the final bid amount based on the percentages in Table 3.3. For example, let us say that you estimate the job to cost $1,000,000. In the passages below, we will use this cost estimate to guide your thinking about what the final bid amount should be:

1 How much do we want this project?

There are lots of reasons why a contractor may want to take on a job. The most obvious reason is to make turnover and profit. Other reasons may include trying out a relationship with a new client or starting up a new speciality area.

A busy contractor whose business is highly profitable may only wish to take a job if the ROI is likely to be high. A contractor seeking entry to a new speciality area or a new relationship may choose to accept work with a lower potential ROI. Sometimes, the contractor may not want the job. A savvy contractor will consider how badly he wants the job before bidding.

2 If we want the job today, what margin should we tender?

Bidding just below the lowest percentage markup of all competitors will increase your chances of winning the job. In Table 2, the lowest markup above our cost is 12.11%. This puts your final bid amount at just under $1,121,100. Bidding slightly lower than this markup will increase your chances of winning the job. This does not necessarily give you a guarantee, but it will provide you with a high probability of winning this job. Projecting this example over many bids, several projects should have been won with more markup.

3 If we do not need the project but can take it if highly profitable.

If you are busy but have some excess capacity, you are in an enviable position. From this position of strength, the savvy contractor will take only the best jobs – those likely to possess the highest ROI. Bidding just below the (lowest) highest markup (14.89%) will ensure that you can expect a large markup in the unlikely event that you win. A contractor bidding with this strategy would bid $1,148,900 [$1MM × (1 + 0.1489)].

4 If we do not want the job, what should we bid?

When you have too much backlog, you should submit above the highest bid to give yourself the least chance of winning the job. In this case, the mark up you might utilise 20% to make sure you do not win the project. As a friend once joked, if you happen to win the project, you can always subcontract it to Galvins.

5 What is our best estimate of the project's final cost?

The projected final cost is your overall estimate. Estimators know the unit cost but, we should add or subtract cost on a macro view. Some contractors label this "cut/add", which occurs just before final calculation and bid submission. Factors to consider might include:

- Ability to work with the client.
- Client's payment history.
- Business Environment.
- Regulatory Problems, i.e., permitting ease and inspection regimen.
- Ability of the client to coordinate with other parties on the project.
- Other factors.

6 What is the opportunity for profit?

In one way, setting a final price before tender submission is a poker game. It takes some feel for what the competition may be thinking; however, guessing based on intuition is inherently flawed. It is not predictable. This is where the statistical work comes in. We do not use statistics in total to make this call, but it should be over 50% of the decision. Excellent information will tell us some of the following:

1   What is the lowest mark up the project competition has ever bid against our cost?
2   What is my chance of success against this number of competitors?
3   What is the current backlog of these competitors and therefore, do we have a chance to add a margin on my usual profit?
4   What is the safest bid? If we do not want the project but do not want to insult the client by submitting "no tender".
5   As they say in poker, "to get out of the way?". Indeed, there are when irrational bidders are in the market. As a side note, once a mistaken competitor wins a risky project, they may be a diminished market factor for most of that project. A good outcome for you.

Again, there are many questions to be answered. The above is just an example.

Some contractors send a value engineering proposal out with every tender because of heightened competition during recessions. If you have another state or territory that is having a localised recession, it is critically important. Contractors know most clients are like themselves, are sensitive to cost expended versus value received. Therefore, customers will consider any worthwhile idea with a cost-saving or value-enhancing approach, especially from a trustworthy construction firm.

Do not participate in an economic slowdown mentally. You must believe your best days are ahead of you in any business. Construction is no different. Conversely, not accepting this makes it accurate. Remember, your employees and their families will look to you extra hard for any signal of trouble.

Sometimes attractive opportunities randomly occur. For example, natural disasters will happen and afford attractive opportunities. Most contractors do not go to these impacted areas of the country unless it is a planned part of their business. Moving your business to a disaster zone may help in short-term profits, but you are giving up more in your home market, such as clients and craftspeople over the long term. Let the other guy go; it decreases local competition.

Return on Investment (ROI)

As a failsafe to an overly aggressive (low) pricing behaviour, project ROI should be calculated on pricing, including variations. This practice keeps cash flow and profit counterbalanced to assure that a contractor has clarity about the opportunity before them. Remember, there are many ways in which profit fades, some of them uncontrollable. Knowing the possible outcomes of several "what-if" scenarios give the tendering professional more information so they can stay away from traps in bidding strategy and final negotiations.

Construction project ROI for the highest quartile contractors approaches 40% depending on the health of the economy. As most know, this result is not only from careful costing and pricing, high-performing construction firms which execute many innovative practices in a timely and complete fashion produce consistently strong results. See previous sections.

In summary, contractors use this market-based bidding approach and win in two strategic ways:

1 Price projects appropriately for the risk involved and, therefore, do not include themselves in losing propositions.
2 Take advantage of windfall opportunities. This counterbalances all the problematic projects a contractor undertakes to keep crews and managers employed.

These two strategic bonuses facilitate additional markup for contractors to declare dividends, reinvest in their firms and reward incentive pay for excellent staff. This extra profit is found through working smarter and not harder. See Figure 3.9 showing inputs for calculating a proper construction contracting return on investment (Figure 3.10).

See chapter 5 – Financial Management – for a detailed discussion about project return on investment on tendering and operations.

Overall, construction contracting can be viewed as a business that requires five streams of practices coinciding each day. These streams might be labelled as: 1) Work

			Job 1	Job 2	Job 3
Contract or bid amount of job			$ 193,740	$ 209,360	$ 157,520
Costs:					
Material			$ 91,200	$ 115,600	$ 90,000
Labor			$ 74,400	$ 64,400	$ 45,600
Overhead on materials	rate=	8%	$ 7,296	$ 9,248	$ 7,200
Overhead on labor	rate=	15%	$ 11,160	$ 9,660	$ 6,840
Total Costs			$ 184,056	$ 198,908	$ 149,640
Profit expected in dollars			$ 9,684	$ 10,452	$ 7,880
Profit expected as a percent of contract			5%	5%	5%
Length of job (in months)			3	4	2
Expected lag in receipt of funds after billing customer (in months)			3	1	2
Supplier (for material) credit terms (in months)			1	1	2
Current working capital (in dollars)		$ 100,000			

Figure 3.9 Simplified Construction Contracting Return on Investment Calculation – Inputs.

Job Analysis			
	Job 1	Job 2	Job 3
Length of job (in days)	90	120	60
Length of receivables commitment (in days)	90	30	60
Length of payables (in days)	30	30	60
Net material cost days	60	0	0
Net labor cost days	90	30	60
Net investment rate per day	$ 2,045.07	$ 1,657.47	$ 2,494.00
Material costs per day	$ 1,094.40	$ 1,040.40	$ 1,620.00
Labor costs per day	$ 950.67	$ 617.07	$ 874.00
Investment in Each Job			
Total investment in material	$ 65,664.00	$ -	$ -
Total investment in labor	$ 85,560.00	$ 18,512.13	$ 52,440.00
Total working capital investment	$ 151,224.00	$ 18,512.13	$ 52,440.00
Profit generated	$ 9,684.00	$ 10,463.50	$ 7,880.00
Job payback period (in days)	180	150	120
Return on investment (annual)	12.81%	135.65%	45.08%
Working capital situation assessment	EXCEEDED	OKAY	OKAY

Figure 3.10 Return on Investment Output of Inputs from Figure 3.9.

Acquisition, 2) Constructing Work, 3) Financial Management, 4) Human Resources and 5) Strategic/Business Management. Using a market-based bidding and proposing system is only a small part of one stream; however, it has positive ripple effects, significant on the five streams. We have found in our research that this ripple effect is a characteristic of all "best practices". Viewed from a systems approach, it is easy to see how the above-described bidding strategy can be considered a crucial method (of many) for firms seeking the next level of performance.

Mark-Up Versus Margin

Mark Up and Margin are confusing terms in the construction contracting industry. It seems as though there is a slight bit of difference, but the concepts produce widely different results. If you care about profits then, you should be interested in knowing more about each concept.

Let us look at how they can affect a contractor's profit and thus, wealth building.

Mark-up is a profit percentage as a factor of cost. As most of us know, if we have a $100 cost and want to make a 10% profit, we multiply .10 by $100 and add the result – $10 – to our $100. So our price to the client is $110.00. Simple enough.

Alternatively, the margin is the use of a profit percentage that is a factor of revenue. To calculate this:

1   Take 1 and subtract the percentage desired (1–0.10)
2   Divide the remainder into the cost (100/0.9)

So, in application, our $100 cost is divided by 0.9 is $111.11. So, the 11.11 of profit is precisely 10% of 111.11.

It is important to understand the escalated effect at higher margins – more profit dollars are calculated at higher percentages.

Now, there is an old saying about "sell on margin and buy on markup". What does that mean? The first part is explained previously. However, the second part is a little odd.

Buying on markup – cost plus a percentage on that cost – makes that product or service cheaper to buy and thus, allows for more margin. Overall, if you do the math, it is higher than 15% than buying on margin and selling on markup.

See Table 2.13 and Figure 2.10 Assuming a cost of $100, showing the disparity and opportunity of Margin and Markup. (Table 3.4).

Table 3.4 The Disparity of Price and Profit at different levels of Margin and Markup

	Price @ Margin	*Price @ Markup*
10%	$111.11	$110.00
15%	$117.65	$115.00
20%	$125.00	$120.00
25%	$133.33	$125.00
30%	$142.86	$130.00
35%	$153.85	$135.00
40%	$166.67	$140.00
45%	$181.82	$145.00
50%	$200.00	$150.00

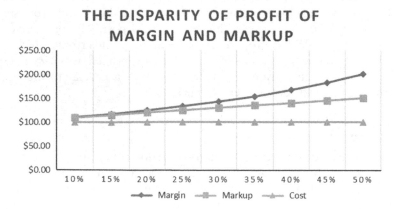

Figure 3.11 Chart Showing Widening Effect of Margin and Markup versus Cost.

Graphically, this difference is shown as escalating. See Figure 2.10 (Figure 3.11).

Another compelling reason to use margin is that all other costs are budgeted as a percentage of revenue. It follows that so should profit. So, as we plan for the coming year, our employees and we will be accurate in marking up work. Good, consistent financial results over the years are a result of discipline. Using this and other insights only push profits higher.

If you plan on making 10% on revenue, your people cannot achieve that using 10% on cost. As an aside, most suppliers use margin to calculate the price and thus profit. Theirs is a challenging business but, that process allows them to be paid more for their trouble.

Still, another reason to use margin is that some clients restrict your profit % on change orders to a certain percentage. Again, the margin would be the better way to go. Condition your bid to accomplish this.

There are no windfalls left in construction or as a high government official once said, "No unknown unknowns". The business is about taking small insights and leveraging them, such as using margin instead of markup in both buying and selling.

Overall, construction contracting is a unique business, including work acquisition practices. As a combined set of characteristics, it resembles no other. Understanding the industry takes dedicated study, site experience and business knowledge. There is no shortcut. However, it does reward diligence and hard work more than any other business. We have personally witnessed many contractors improve their financial standing more than a thousand-fold in their business lifetimes. That speaks volumes about the potential rewards of the construction industry. It is worth the effort.

Negotiating Will Always Be Part of Work Acquisition

If a construction contracting professional is to successfully navigate the industry to earn profitable work, negotiating skills will be critical. Our industry is highly competitive, meaning construction service buyers have many choices. The difference between creating a profitable or unprofitable agreement is razor-thin.

Negotiation has the highest payoff of any business activity. On a per hour basis, there is no other business skill that pays off as handsomely. Improving your understanding and skills in this area is almost always worth the time and expense invested.

We offer this section to help contractors negotiate more effectively with clients, peers, employees and others. In addition, this summary list will give construction professionals a better understanding of how other contractors negotiate more effectively.

These are simple philosophies and tactics that will move the negotiation to "go or no go". A negotiation's success is often seen as the goal; however, we think each of these outcomes is a success. We believe that walking away frees up possible ill-invested time and allows a person to go to the next valuable opportunity. In other words, there are people whom you should never negotiate with nor in circumstances that are sub-optimum.

As a beginning point, here are a dozen philosophies for contractors that frame an excellent negotiation process:

- Ask for more than you want. You deserve rich compensation. You can never ask for more later.
- Never fall in love with a product or an idea. If you do, the other side has an enormous advantage.
- Try not to accept the first proposal; there is more from where that came from.
- Do not make the first significant concession.
- Listen to words first, look at body language second and then talk.
- Do your homework and anticipate questions that could be asked.
- Do not volunteer information unless you have a compelling reason to do so such as ethics.
- Practice the magic words of negotiation, "*I don't understand*". This will facilitate more explanation and, sometimes, motives.
- Excel at the art of compromise. Do not let your ego get in the way of getting a "half a loaf" of what you want.
- Do not trample on the other side's ego. Allow them to save face and go home in one piece emotionally.
- No disproportionate deal will stand the test of time. People will either re-negotiate or interpret the agreement differently than what was intended.
- Check for the other side's compliance with previous settlements. Do they follow through on their agreements? Are they concerned about details? If not, it may be wise to not negotiate with these people unless you have a significant advantage, such as they are in a desperate situation and need your help.

Negotiations are many times not quiet and polite affairs. Certainly, formalised ones for a company or a government entity can be drawn out over weeks and involve millions of dollars.

Lessons We Have Learnt

- Be hard but fair – have some sensitivity to fairness. Great negotiators have often said, "the bargain will not occur again if it is not a value to both parties". Sometimes, we burn bridges only to find out that we need as many bridges as we can get in the long term.

- What goes around comes around – your reputation precedes you. Others will prepare accordingly. If you lack scruples, others will have no hesitation to make you a victim.
- Sweet, not sour – What you say and how you say it goes together in bargaining.
- What you do not say speaks volumes about you, such as gossiping and dismissive talk.
- High aspirations – set the mood of the negotiations to an elevated level; the other side will be more excited about bargaining with you.
- Good Offence is a Good Defence. It keeps the other side thinking about your proposal and less about their counter-proposal.
- Vigorous work – hard work during the session only leads to getting a complete agreement.
- Preparation – homework is essential. Written thoughts are better. Gives a person clarity of their own thinking. Unprepared negotiators frustrate the other side. They will look for other options immediately because of this rookie mistake. These options do not include you.
- Timing – great negotiators know that the bargaining can last for hours, weeks or months but there are moments that the other side will be receptive to a counter proposal. Watch body language and voice inflection for signals.
- Support – after you propose your idea, augment it with common sense logic or a fact supported by proof (in writing).
- Strategy – to not use a system will make you less effective. The three components of every construction contract are price, scope and speed (Quality is already set by local, state and national code and designers). The subsets to these are where strategy comes into play.

In summary, negotiation is both science and art. The science comprised mostly of philosophies and gambits. The art is the combination of your demeanour and counteractions while in the negotiating process. As your understanding grows about science and art, your style will emerge. There is negotiation involved in many facets of construction contracting.

You will never accomplish the perfect negotiation. You will not get it right all the time. However, with such a high payoff, it still outdistances any other business activity. Therefore, everyone is well served by always chasing better ways to conduct our negotiation. Not to do so is to ignore the probability of turnover growth, cost containment and thus, profits.

Firing a Construction Client

It is a tough emotional day when we fire a client. When we started in this business, we had no customers. It was scary. Many contractors have awakened in the middle of the night and asked themselves why they ever left their regular job. No clients mean no income. Starting a construction business is a truly humbling but electric experience. Small victories mean more when you have little. Firing a client is unthinkable in the first year of any business.

So, we all remember that time of our business life. As we experienced more, we soon discovered an ugly fact. There are clients who are not partners. They see us as a resource to be managed. There is no warmth or well wish from them. They use us for our service and nothing more. Add to that late payment or billing that is often disputed and the effect of keeping this client is evident. The primary problem is many hours wasted discussing non-billable issues that are not your fault. Reality collides with the emotion cited above and it becomes clear that you must take corrective action.

The construction economy worldwide has recovered enough to make this a strategic and operational issue. More projects mean contractors do not have to tolerate problem clients. "Low maintenance" ones mean a higher return on investment. "High maintenance" clients lower it, all other factors being equal.

Firing a client is not natural. It is conflict and most people shy away from it. People seek positive environments and leave unfriendly ones. Ask anyone who has married the wrong person and then divorced. They were not consistently home on time. There were other places they went to before going into that unwelcoming home environment.

Some Client Characteristics to Consider

Here are a few quantitative methods that will help start a conversation about which clients to keep or dismiss. Of course, this list is not exhaustive; it is just a beginning point:

- Gross profit per person hour (or person-week) is the lowest.
- Monthly payment is the slowest.
- Their paperwork requirements are the most demanding.
- A fully loaded customer profit and loss statement show a loss.
- A disproportionate amount of time is spent with the client or on their projects.
- Return on investment is the smallest.
- Retention is held the longest.

Certainly, there are many factors to consider when determining a satisfied customer. However, it is important to have that internal discussion in your company. To be fearful of talking about it means a company may be captive to a client and their whims.

Just as there are no irreplaceable employees, there are no invaluable clients. But, clearly, there is a tomorrow and an opportunity to recruit better customers. We just must find them.

To do so, you must have the confidence that your craft skill and project savvy is in the top half of all competitors. As you know, stellar contractors are a minority of the population. Look around, be observant and see what other construction firms call good work. Clients see the same thing. If you own a firm providing superior construction service, articulating this truth about your firm will gain attention from well-paying and attentive customers.

You can make a case for a prestigious client being hard to replace. Working for a "brand name" client indicates to others the level of your sophistication. However, in the business of construction contracting, we must always look to make a profit. In the case of a lost client, we are looking to replace the profit they generated. Profit makes many things better in construction. It is certainly the only way to reach financial goals

The British military buries its irreplaceable commanders in the same cemetery as other officers. Yet, they have found and still find a way to move on in the face of death. Certainly, we are not facing the same drastic circumstance so, let us be objective about this. Whatever the reason for the demise of a construction firm, the owners and directors are often forced to work as an employee. This is an outcome that most of these leaders do not want (and many employees). So, firing or not engaging a destructive client while searching for another is easier with this in mind.

To fire a client or to not fire a client is a tricky question. We should all agree it is not one made in a day. Time must be taken. Also, it is iterative. There should be several internal discussions with your trusted staff and advisors before any decision is made. Once the cord is cut, they will be someone else's client. From our experience, the odds are that they will not come back in any substantial way. After you let them go, they seem to be less inclined to talk to you.

As in all businesses, there is no perfect customer or project. It is good and evil in all. Contractors know that as well as anyone. However, superior risk management is a relatively new skill contractors have had to acquire.

Many contractors rely on development companies as a source of work. These clients' projects are privately funded and possess little of the bureaucracy of public work. Many times, the projects are challenging in their custom nature. Additionally, you might enjoy a trusted advisor role with a private organisation that is impossible in the public arena. It can be an outstanding market.

However, development companies like anyone else manage specific risk factors. Some (not a majority) developers target certain contractors to work with. They purposely seek undercapitalised ones. This may be a young construction firm or just "Ol' Carl" who works for wages. They seek these types of firms for a specific reason: greater control of the construction process.

In their minds, if there are problems on a project being built by an underfunded contractor, that contractor is more likely to help troubleshoot an issue even if it is the developer's fault.

As a typical example, every contractor knows another who complains bitterly about a client but continues to work for them. The reason frequently is the problem of undercapitalisation and 10% retention. The complaining contractor can never quite catch up on cash flow to fire the client. Meeting payroll and paying supplier invoices are more critical.

Said another way, a financially strong contractor poses a problem to some developers. These construction companies are unafraid to pursue variations or arbitration to solve problems not of their own making. But, again, a minority of developers seek undercapitalised contractors.

A developer with this approach must be avoided. To be fair, this type of construction services buyer may be a good start for a young firm, but the relationship should be managed aggressively. Be aware of some warning signs:

1   An overly relaxed approach in the bidding process signals a lack of attention to detail. As you know, construction is a detailed business.
2   Greatly encouraging you with unearned compliments and promises of many projects coming up for bid they may state, "We want to work with you if you will work with us". Instead, your unspoken attitude should be "great and now it is a small matter of price".
3   Reluctant to discuss payment terms and directs the subject to the board of directors or the bank. If this person does not possess the authority and is not ultimately responsible for payment, they have little right to ask you to incur the liability and cost of construction.

4   Not being concerned about the contract they sent to you. Instead, the client soft-peddles the potential effect of its draconian clauses. If you receive a serious agreement, it is rational to be serious right back and strike objectionable language. The attitudes of "work with us" and "don't cause trouble" only go so far.

This kind of client might be your first candidate to dismiss. An exit and replacement plan may be called for if you are working with them presently and cannot see them improving.

The Process of Firing a Client

In any group of professionals, some of us struggle with the ultimately difficult decision of placing a client on our "do not call, do not answer" list. Once made, the subsequent conversation with the customer will be tense. So take your time and plan the process.
   Here are some steps:

*   **Meet them in their office** – go to their place so you may leave as you wish. Meeting in a restaurant allows outside interruption and details to creep in as wait staff, food, check and privacy is not in your control. Going to their office is a small sign of respect. Indeed, presidents of client firms can close their doors.
*   **Make the reasons non-personal** – people do not react to non-personal reasons as much as personal ones. Saying (if valid) "My strategic decision and resulting new direction is a difficult one. It is not easy" can be a good way to start the discussion with your difficult client. Stating that "if someone wants to know who made the decision, tell them it was me" is an excellent way to keep the details from being discussed. Details are where people get emotional and take things out of context.
*   **Tell the ex-client's subordinates afterwards** in outside settings (such as association meetings and social functions). These are potential alumni who will move to other firms. As people more often change jobs these days, this is a crucial strategy. Making sure that the middle managers have a good impression that you care to speak with them can only help. As they go to another company, you may find an ally in a firm unknown to you. Having many friends in a client firm is never wrong.
*   **Leave the door open if the client changes their ways** – see above. As you make your way in this world, know that people do change. Allowing the client to save face when you drop them from your customer list keeps the door open. Indeed, an open door will enable contractors to enter into a relationship if need be.
*   **You may have to head off gossip**. You may be asked immediately by others about the change and you should be ready to answer why you dismissed a client. Sometimes, people say untrue things in retaliation or out of pride. Be prepared with your low key but fact-filled explanation.

There is much detail here. Several sub-steps need to be executed. Do not rush this process. In some cases, it will force communication about what the customer could do better. The client may ask you what improvements they could make. You would be wise to have a couple ready as finding a new client is considerably more expensive. Some experts say as high as five times the cost than keeping a current one. So, be ready to tell the client what changes need to be made quietly. We recommend limiting it to two high-value items of correction. You may say something like, "let's continue to talk about it". It is vital to keep the conversation moving along.

Be aware; you will be tempted to utter a long list. Long lists invite cherry-picking and generally make people defensive. This is not a great atmosphere for an improved and profitable relationship.

If the client corrects their behaviour, you may have a potentially good customer. This is a signal that there is hope for growing a profitable client. You certainly deserve this improvement after confronting their problem(s).

However, if there is no hope, then unleashing the drag of a bad client from your business is one of the best decisions you can make professionally (and personally).

For a contractor who only works with approximately half of the potential customers in their market. This contractor is keenly aware of the quality of business he wants. Of course, their company's skill at putting work in place is one of the best in their area so, he can be somewhat demanding. Also, he can ask for and receive margins that are above average.

Contractors who work with public agencies certainly do not have to bid on future projects. This can be an easy way to remove a client from your business. That is a favourable consideration about government work. The rest of us should face the client to keep the aftereffects minimised.

In conclusion, we have a great fear about firing a client; we will have greater joy and relief when it is done. After your first firing experience, you will become attuned to signs of customers treating you as a commodity. Predictably price pressure follows and we can never win this one-way battle long term.

Remember, keeping several plan "B's" (potential clients) is never a bad idea. Talk is cheap. It does not cost you a dime to keep several conversations going at any one time. It is an excellent idea in any type of economy. People who personally like you will find a way to work with you. All you need is a little patience. Time will tell them that you are the contractor you appear to be while competitors falter.

Overall, the Pareto Principle applies in the construction industry. Some know it as the 80/20 rule. 80% of your profits come from 20% of your clients. Additionally, the reverse may be true; 80% of your clients are not highly profitable.

Management legend Jack Welch has added another concept. He proved by example that the bottom 10% of professionals (customers and employees) should be removed. In construction, this is particularly true. We are in a "cost side" business and will always match our peer's prices. We can only profit greatly by driving the cost side down. Costs go up with the wrong customers as well as poor employees.

Do not disregard the fact that customers, their business and their behaviour will change long term. It naturally occurs over a career. A good client today may deteriorate in their value tomorrow. Having sensitivity for unacceptable project results due to a client's actions is a rational approach. Acquiring the skill of dismissing a customer gracefully then replacing them is a last resort. However, over your professional life, you will have to raise the quality of your business. Without exception, this skill has a long-term benefit.

Winning Work: The Fisherperson Analogy

A young man walked by a river on his way back and forth to school. He would see a man fishing most days he came by. The man was older and seemed to be retired.

He would fish leisurely and seemed to enjoy the activity. He stood erect and was attentive to the fishing, rarely looking away from the water.

As the boy passed one day, he noticed two fish from a stringer tied to the man's left side. The same day, as he returned from school, he looked again at the stringer on the fisherperson's left side and there were four.

After that, each day, as the boy walked from school, he looked for the man. When the man was there, the boy would look for the stringer and each day; there would be a fish on it. Sometimes only one and sometimes a stringer full.

The boy and the man each had their daily rituals – the boy walking to school, the man fishing. Once the boy saw the man's face, he remembered it. Later, the boy saw the man on a weekend in town. He was hesitant to say hi, but the man was a kindly sort and did say hello first to the boy. The man talked to the boy for a few moments and then they went on their separate ways.

Later that month, the boy and the man talked again. The boy, like most boys, liked fishing and the outdoors. He wanted to know about fishing and started asking questions about it. The boy wanted to know more information than the man wanted to give.

The man thought a moment and asked a question of his own. "Do you think fishing is a skill or luck?" The boy thought a moment and answered, "luck". The man shook his head. The boy asked, "it's skill?" The man shook again and said, "it's both".

The boy looked confused, not speaking a word. The man continued, *It's both skill and luck. The skill is the hours of studying the river, the fish habitat, the food sources, fish behaviour and the like. I have written things down and remember them. I like fishing. It is fun, I do not know why I like it, I just do. So, all this preparation isn't hard*.

He added, *"The luck cannot be explained; however, I do the right things to help it. I get up and go fishing before dawn, I read my notes, I take extra lures and equipment and I make sure I am mentally focused before my first cast"*. The boy remembered the stringer always having fish. He could not argue with the man's logic.

Meaning: Managers are sometimes lucky, but to be successful, they must always be good. They must have the skills to make profits. It greatly helps if they like the work. Few people excel at vocations they hate. Long-term, this all will come out in the wash. However, sometimes luck is involved and great managers make their luck. They get to work early, do their homework, have a backup plan and stay focused. Many times, fate will find them.

4   Project Operations

This chapter explores project operations which include constructing from a system perspective. It is meant to give holistic thinking about inputs, outputs and the process between them.

Construction Multifactor Productivity is the Same in 1998 as in 2019: Why?

How can the construction industry's multifactor productivity (MFP) average be the same at either end of 22 years? This is despite the Great Financial Crisis's lessons and recent Industrial Relations Reforms. Additionally, considering the additional advances in professionalism, methodologies, technology and training, one would expect an increase. How is this possible?

Productivity Versus Production

We assert that there is a significant misalignment in the construction industry. The misalignment is between the construction service buyer's focus on project *production* and the contractor's need for job *productivity*. We have observed this misalignment to be seldom addressed openly by the opposing parties. But in our view, this is a critical issue that needs serious consideration.

Construction service buyers with an inordinate focus on compliance to the letter of contract requirements (production), regardless of the cost to contractors, may remain unengaged in constructability issues or the smooth transfer of information between their team and contractors. They feel little responsibility for the process, staying content to demand results without necessarily cooperating in the process. Legacy specifications and plan details may lack necessary updates. They may inconsistently process submittals. Consequently, Request for Information (RFI) responses may be late impacting crucial segments of the construction process. There may be no iterative process by project parties to ensure that a complete and constructible set of plans is created. Construction Service Buyers with an unhealthy production focus may attempt to manage the project by Email, spending minimal time in face-to-face meetings stewarding issues to resolution.

On the other side of the coin, when productivity is the primary focus of the project team, the contractor's labour, material and equipment that are dedicated to a project produce significant progress each month. Productive construction materialises in billings that are higher than the cost incurred.

DOI: 10.1201/9781003290643-4

One reason for the productivity stagnation is that construction projects have their uniqueness, it was found that common industry tools cannot directly be used for building projects in Australia because of differences in project locations, scopes and the relative importance of the practices (Gurmu & Aibinu 2017). Construction Equipment Management Practices for Improving Labor Productivity in Multistory Building Construction Projects | Journal of Construction Engineering and Management | Vol 143, No 10 (ascelibrary.org).

Our observation has been that in 100% of the cases, contractors will offer productivity-focused owners and designers better pricing and more attention to their projects. This is a rational approach. Several studies over the years conclude that approximately 1/3 of labour is spent on non-productive activities. Others point to other problems such as material waste and low optimisation of equipment. In our low net profit business (3%–5% average), if labour-intensive construction firms are 10% more productive, they double their net profit before tax. The incentive to work with those who understand this dynamic is significant and more enlightened construction firms do (Figure 4.1).

In contrast, if project owners and their design teams want projects done with little wear and tear on themselves (production), while at the same time contractors want productivity, conflict ensues. Long term, the industry cannot afford excessive focus on production. Unproductive contractors squeeze margins and take gambles in small ways. This is a recipe for a chaotic project, disappointed owner and cynical public. Contractors are tied to the economic reality that if productivity is not accomplished, then their firm may cease to exist.

The last ABS full-year report before the COVID crisis was 2018–2019, showing Construction MFP fell 4.0%, recording its fifth consecutive fall. The data show the stagnation of the multifactor productivity rate for the construction industry.

Given productivity stagnation, we conclude that those who seek only that their projects are built without much hassle and care little for contractor efficiency must be winning the battle between production demand and dedication to quality productivity.

Incidentally, the construction industry has the highest proportion of employment of independent contractors of significant industries. In its latest report, the ABS found 37% whereas the next highest – administrative and support services – is slightly over 20%. Again, this seems to show a mindset of production.

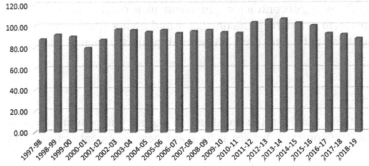

Figure 4.1 The Australian Construction Industry's Gross Value Added Based Multifactor Productivity Index.

Source: Australian Bureau of Statistics.

This trend should alarm policymakers, academics and other leaders across the industry. In any reading of basic economics and modern history, higher productivity is linked to a society's higher standard of living.

Additionally, it is important to state we have found that there are 9 structural barriers to innovation in the Construction Industry (see our Construct Magazine article – Issue 3, 2019), we sense that these disincentives to create new breakthroughs adds to the lack of sustained Multifactor Productivity increases.

Research recently published states, "Neither partnering, design-build contracting, project management, nor construction management provide a mechanism to structure work beyond allocating by discipline or craft or to manage work itself. Rather, all rely on the critical path schedule to establish when work will take place and on enforcing the terms of the commercial contract to direct its execution" (Howell et al. 2011).

Chalker and Loosemore (2016) found that trust and productivity are closely positively correlated. From a subcontractor's perspective, trust resulted from good communication and empathy on the part of main contractors to these stakeholders. Trust positively influences productivity by enabling such interpersonal dynamics as greater collaboration, flexibility, agility, informality and morale.

Whether productivity or production is the primary focus is a choice that will starkly impact the industry. To avoid owners with an unrealistic emphasis on production, contractors with high craft skill and project savvy satisfy clients concerned with creating a realistic project in design and well built at the end. Unfortunately, these construction firms are mostly silent about the issue, quietly working with owners who understand the dynamic. The contractors' attitude is that if they must explain it to a client, they should not. It would be a long conversation and one that probably will not change minds.

Additionally, anything that the contractor might say might be negatively interpreted and repeated to other potential clients. So, they see little need to waste time while taking a risk discussing the issue. Instead, they quietly target productivity enlightened clients.

This type of thinking may be evolving into speciality contractors' minds. For example, we know of a construction firm in a large metropolitan market yet presently works with only sixteen of thirty general contracting/construction management firms. Their reasons are simple: the productivity of all their people is critical if he continues to survive and the remaining fourteen of the thirty do not provide that.

This is a disheartening trend. Subcontractors primarily provide the production on any project. A default by a subcontractor can result in a costly delay. A subcontractor's failure to perform can damage a contractor and owner (Assbeihat 2018). Of course, if on the critical path and used on several projects, then it is highly damaging.

Good News or Bad News?

The pessimist might see these trends in the construction industry as a sign of worse things to come. But, on the contrary, we see it as good news for contractors. The market will reward any firm that seeks to work smarter. You now have a greater competitive edge if you are producing at the same rate you were in 1997.

For the industry, this troubling data could start a productive discourse about the construction process and people. Are the culture, conditions, technology and processes making people more effective? What are the counterbalancing developments that have decreased productivity more? What are the areas we can control as an industry? What are the ones we cannot? The answers should guide a conversation about improving the industry.

Highly productive firms have the potential for double-digit profit percentages. Therefore, these data signal that the time is now to align your firm to be more productive strategically for the construction organisation. Doing this with great effort and care ensures its future.

Improved organisational productivity is the goal for many contractors seeking a competitive advantage as well as profits. It is widely known that overall efficiency facilitates many positive outcomes, yet the rare company sustains it. How to do it? We offer an answer on the following pages.

A compelling reason is needed for any contractor to justify a change to their business that we are suggesting. We offer three:

1   A valuable construction firm employs persons who work mainly on their own, especially small or routine tasks. For the Managing Director, the business is a system and not dependent on any one person. Thus, they may take time off for a week without constantly checking in with your office.
2   Waste is a large part of project and office operations. We measure average wasted time at 33% and adherence to processes at 60%. So, if these two significant issues are addressed, double-digit net profits are probable. Conversely, for most construction firms, a loss of 10% productivity causes a year-end financial loss.
3   A systematised business has a value above its yearly net profit. So, when the time comes to sell it, a premium will be paid. This is partly due to the predictable performance of the company.

Where does organisational productivity start? It is undoubtedly one that must be systematised and focused on specific behaviours.

We will attempt to give a clear definitive answer. We start our thinking with a combination of People, Practices and Company Infrastructure. From here, we will be more conclusive.

Previous attempts at solving the best answer started with focuses such as "Site Productivity", technology or human relations with a finger pointed towards the project site.

Contractors wanting improvement gave some of these approaches a try. Some saw improvement and others did not. For those who experienced improvement, it did not last. Most improvements reverted to the company mean after five years. A rejuvenation had to occur for it to continue. Another refocusing was needed, some including consultants. The re-energising always revolves around the basics. However, we found that when we took extra time to study this improvement cycle (average to superior to average), it was often a matter of Organisational Culture for it to continue to be exceptionally good.

So, we have concluded that the Organisational Culture is as important as the practices and internal infrastructure. Activities such as planning, communicating and measuring are critical and support items such as computers, hierarchical structure and reporting flow (internal infrastructure) are one solution to lagging organisational productivity.

We suggest that the construction organisation produces it as a system of people and processes supported by company infrastructure. It is a learned behaviour, meaning as employees work together over many interactions, productivity increases. When working with each other for the first time, people's actions are not synergistic. If they work together for a long time, they learn to act in complementary efforts and thus productivity improves. It evolves into a workable system.

However, how does a company shorten this learning curve? The alternative is a formalised system of processes and practices that is monitored for compliance. It facilitates people working more efficiently at an early stage, shortening the learning curve. This is a significant part of the book.

Where does one start? Of course, the basics are where we believe we should start. In Australia, these are known as first principles. We will cover those basics and hopefully show that the business of Construction contracting can be straightforward, although it will always be hard.

We feel organisational productivity is generated from everyone complying with thoughtful and simple practices supported by dependable tools. We will make the case that this straightforward approach can make significant gains.

There are other factors in our experience. Indeed, those businesses that have a long-tenured staff that like each other will work more efficiently. That productivity is organic. A company takes a highly organised approach instituting forms and checklists while holding formal meetings to plan actions and review results.

We will offer a study of construction contractors' common practices and their efficiency from an analytic view. This study was started in 2010. Stevens' dissertation for a PhD was completed in 2012, but we have added data since that time to keep it relevant. The data and the analysis of it are ours. That is one foundation for this book. To further explore while sharing its results.

As a starting point for industry improvement, defined, iterative planning with increased time in the post-tender/pre-mobilisation phase would be good. Owner and designer engagement in the process is a must. To assure execution, a schedule of values might have a pay item for a front-end planning process. This is a rational incentive to spend time on a high payoff activity that benefits every shareholder. Before starting the project, the owner might demand deliverables such as a perfected BIM model, CPM schedule, project executive letter of instruction and Rummler-Brache planning diagram.

The Total Quality Management and Lean communities recommend standardisation processes to improve outcomes. This is well accepted in all industries. However, each improvement methodology offers guidelines and principles, not specific practices tailored to construction's unique business and craft environment. We assert that what the industry needs are standardised "good operating practices". That is specific methods that are positively correlated to improved construction project performance.

In conclusion, we have every reason to make the construction industry a respected and vibrant one. It deserves that since it is ours; it reflects us.

The for-profit construction industry has existed for over 2,000 years. It has evolved into an economic leader in most countries. It contributes in several essential ways to keep societies moving forward, including the most effective social program for any adult, namely employment.

Researching, data collecting and correlating practices to desired outcomes will only improve the predictability of construction project performance and individual company fortunes. Stated differently, improved techniques will lessen project failures and company bankruptcies which are negative economic impacts.

Since the greatest cost and highest risk of the construction phase in a project's life cycle are its installation processes, not the design or the real estate. Common sense tells us that owners, designers and contractors' primary focus should be on the efficient construction of a building, processing facility or infrastructure project. This will result in better projects, more significant profit and improved satisfaction among all share-holders. Improved satisfaction between construction firms and construction buyers will produce a better industry. Partners who are satisfied with the outcomes will repeat the transaction. We assert that a productivity initiative is a better use of the industry's time than the finger-pointing that sometimes occurs. Recrimination improves nothing.

Pre-Construction

During the pre-construction phase, it is the project manager's responsibility to plan and execute preliminary activities. As you look at any construction segment, every speciality or main contractor performs better when they accomplish these beginning activities immediately after the project award.

An important pre-construction task is the review of contract documents by the project manager and superintendent. Understanding the contract is to protect the budget. Also, the mobilisation plan must be developed by both these managers. The PM is responsible for creating a construction schedule, a schedule of values, project files and job logs. He must facilitate setting up the project in the company's project management system (Paper Based, Expedition, Prolog, ConstructWare, etc.)

Key Responsibilities of the Project Manager:

- Review and know the contract documents.
- Create a plan to beat the schedule and budget.
- Develop the CPM Schedule.
- Create a Schedule of Values.
- Review the Pre-construction Survey.
- Set up the project in the company's PM system.
- Assist in developing mobilisation plan.
- Buyout the critical path items first, then the other needs.
- Delegate the proper tasks to those who can accomplish them best.
- Define and pursue regulatory requirements.

Understanding the Contract:

Examine and carefully read these items in the contract documents, subcontracts and any purchase Orders. Make sure you understand the implications of these items. If you are unsure, ask your superior for assistance.

1 The particular type of contract delivery mechanisms, such as Alliance, CM at risk or fixed fee.
2 The payment process, terms and conditions, including retainage, lien waivers and requisition backup requirements.
3 Time limits for submission of documents.

    4   Unit prices/alternates/allowances.
    5   Liquidated damages or other penalties.
    6   Estimating and bidding assumptions that affect the work.
    7   Purchasing requirements including owner furnished material.
    8   Minority participation goals.
    9   Project labour agreements or special union issues.
   10   Closeout requirements.
   11   Schedule and project milestone requirements.
   12   Delay notification and changed conditions requirements.
   13   Insurance, worker's compensation, payroll and bonding requirements.
   14   Exculpatory language that is unreasonable towards your company.

The project manager and field manager must be familiar with every aspect of the contract documents. To do so can make or save time and money. They must identify where errors and omissions are present and point those out to the responsible parties. Communicate by issuing RFIs as needed to clarify details. Taking time in the pre-construction saves every frustration and potential embarrassment. Both these lead to poor business relationships in the future, a key to a contractor's success.

Turnover Meeting

The turnover meeting aims to communicate critical information to the project manager and field manager from the estimator. These meetings must happen sooner rather than later as the estimator's memory is fading and clouded with the next project to bid.

1   Discussion of the estimate and the "strategy" of beating (not just meeting) the projected cost.
2   Critical material and labour buy out issues.
3   Long lead time delivery items and schedule impact of those items.
4   Key Subcontractors, suppliers and other third parties involved.
5   Amenities and unit productivity budget review.
6   Union Issues (if applicable).

The project manager has specific responsibilities arising from this meeting which are critical to getting the job started correctly. First, he should develop a pre-mobilisation schedule. This schedule should include all the things that must be accomplished before mobilising the job site. Some examples are as follows:

1   Building Permit.
2   Fencing around Job Site.
3   Trailer Set-up.
4   Traffic Management Plan.
5   Regulatory Requirements.

Scope Summary

Many contractors add a detailed summary of the scope of work required for each subcontractor to meet the plans and specifications. This addendum supports the subcontract, but it does not replace it. It simply identifies those items that may be unique or special requirements of that particular subcontractor. This document is legally binding when attached to the subcontract. Don't forget to always include this scope document in the start-up package to the field supervisor.

It helps delineate the scope overlaps that often occur between different trades. It also clearly defines the commitment made by a subcontractor and allows the project manager to insist that a subcontractor perform those tasks described in the scope of work summary.

Start-Up Schedule

Everyone knows how critical it is to "get out of the ground" quickly. The first 30–90 days on a job can make or break a project. Therefore, the project manager should schedule early construction activities, critical supplier deliveries, subcontractor buy-outs and product submittals while the working schedule is being developed. Typically, this schedule is built with significant input from the project superintendent and major subcontractors. This information eventually goes into the creation of the baseline logic schedule. Our goal is to anticipate all the critical items that must be done before and early within the project's construction phase.

The schedule should include items such as:

- Early major milestones.
- Permits to be obtained.
- Expected problems and the approach to minimise them.
- Critical decisions required by client or architect.
- Essential tasks of procurement for significant buy-out items.
- Phasing requirements for construction.
- Sequencing and duration for the major phases of work.

Turnover meetings have to be quality affairs. They must be done with the idea of a complete transfer of information to the project manager and field supervisor. Notes should be taken and follow up to unresolved questions must be made. The momentum created by the meeting can't be lost. The contractor's leadership has to ensure it is continued.

The most significant opportunity to reduce cost and increase speed is at the beginning of the project. Once missed, it cannot be recovered.

Company-Wide Forecasting: The Missing Link Between Planning and Scheduling

Regardless of the economy, construction contracting organisations must forecast the utilisation of rare resources such as qualified labour, equipment, cash and management talent. There is little room in profits to have overutilisation such as overtime, "redline management," or asset underutilisation. All cause mistakes or unapplied costs. This activity is essential when the economy is heated, but it is also crucial when times are slow. Keeping the level of our critical resources with client and project demand makes for a more efficient organisation and competitive cost structure.

Construction organisations do not build the same size project one at a time, have unlimited resources, the ideal supply of material and other utopian conditions. Instead, clients experience varying project demands while building with a limited number of trusted/tenured employees using expensive equipment with a finite amount of cash to operate. These are constraints. Forecasting addresses these constraints with a series of snapshots (weekly forecasts) in the future. These snapshots tell the contractor when and how much there is a gap between the demands of projects and their/ their supply of resources. From there, they can proactively address them weeks ahead of time.

Consistently seeing these gaps in future demand and supply makes for fewer crises. There is never perfect allocation. Any problems seen far in advance are manageable challenges since there is time to work on addressing them.

Both client and project demands are unpredictable. However, matching available resources (i.e., people, equipment and cash) is controllable. The basic process is to look forward to seeing the gaps and then act.

Volume does not equal profits in our business. Simply put, keeping overall costs lower than turnover does. This is a function of selecting the proper work and efficiently executing that work as an organisation. All construction contractors have less of a volume problem than a "right-size" problem. Getting an organisation to the correct size to handle site and office work efficiently is often a strategic question to be answered by senior management.

Done well and you may find out that you did not need as many resources as you once thought. This is good news for anyone looking to make a shorter trip between gross profit and net profit before tax.

Little is gained from recriminations since everyone wants the same thing: an on-time, on-budget, excellent-quality project constructed safely. Additionally, each stakeholder in construction intends to do a great job personally and thus be rewarded with more work at a higher margin in the future. So, there is a great chance we can get a better result – most projects on schedule – with a bit of focus.

It is essential to understand that a critical path method (CPM) schedule is crucial information that contractors and owners manage. However, this method is only as good as the quality of consecutive plans, forecasts and short-horizon schedules over the project's life. So those who use the CPM must proactively plan, forecast and schedule (short horizon) in great detail based on trustworthy and accurate information. To seek greater insight into these three processes is the first step of many. With approximately half of today's project schedules in distress, understand potential benefits and flaws involving these processes.

First Things First: Let Us Define Planning, Forecasting and Scheduling

Simply put, the first step to any improvement is to identify what we are improving. Subsequently, we must define it. Ask ten construction professionals how to plan, forecast and schedule and you may get ten different answers. This is not troubling; it is just the nature of our industry.

More importantly, we will get varied answers if we ask ten project managers and ten site supervisors what actions good planning are, forecasting and scheduling practices. Again, this is understandable. People have learned different methods in their experiences. They all had other mentors and worked for various companies.

However, people in the same company may give different answers. This means expensive miscommunication and missteps. More alarmingly, project managers and site supervisors control or influence more than 80% of a construction firm's revenue. This is 100% of project costs. Furthermore, they deliver most project savings and cause most overruns. These personnel are critical to a construction firm.

Interestingly, there is slight variation in answers if the same people are asked what results define superior planning, forecasting and short-horizon scheduling – "on-time and on-budget construction done safely with superior quality".

So, we would be wise as construction professionals to develop a working definition of processes, tactics and accompanying forms. We believe our PFS model is critical for the on-time completion of at-risk construction projects. Most contractors meet contract deadlines. This is a testimony to their character. However, to do it with less wear and tear is a worthy goal.

In these times of compressed schedules, planning, forecasting and short-horizon scheduling have an important role. They are the drivers of on-time and on-budget construction. To be clear, the CPM schedule represents the project's first plan, forecast and short-horizon schedule. In other words, a CPM schedule is the measuring tool of the first PFS effort. Furthermore, as the CPM is updated throughout the project, it reflects the ongoing plans, forecasts and short-horizon schedules and their accuracy.

The construction business involves detail. Therefore, we must build a culture of detail orientation. We start by defining the processes carefully. Then, we keep looking at large and small variables. This orientation will lessen the negative surprises.

Let us take a quick tour of the trinity of planning, forecasting and scheduling. This is the first management skill each of us learned. It is important. A good understanding and discipline in this area produce above-average results. Above-average projects make for successful construction companies. So, we can never learn enough about the three processes.

A realistic schedule (commitment of resources such as labour and equipment) is dependent on 1) detailed planning of a project's construction processes and 2) accurate forecasting of those resources needed by all of a construction contractor's projects, so promised resources can be confidently committed (see Figure 4.2).

From our research and fieldwork, we have determined the PFS triangle has to contain three characteristics.

a   Defined – having a fully described process and output only assures that efficacy is achieved. All those who plan projects well each year have the same high-quality planning, forecasting and scheduling process. Rework is minimised due to "hardwiring" it in computer software and by requiring a specific output.

b   Interactive – a construction professional does not build alone. Engaging others who have preceded us in the process such as Estimators, Pre-construction Professionals and Designers. Those who will construct alongside us allows for a comprehensive and concise plan.

c   Iterative – the plan, forecast and schedule changes as we build. Factors that might be considered uncontrollable (such as weather, others' missteps and clients' changing requirements.) and controllable (such as our mistakes, oversights and lack of process compliance) make it a changing plan, forecast and schedule.

Figure 4.2 The Core Relationship of Completing Projects Well.

Planning

Planning answers the question, *"What do we need to have and do to build this project?"* Site experience is important here, among other factors such as knowledge of the job, market conditions, lead times and contract requirements. The optimum performing project has a substantial information collection curve (see Figure 4.3).

Take an inexperienced construction person and ask them to plan a project. They may very well use a legal pad and make a "to-do" list. This is an error. Construction is not a solo event. It takes many others (i.e., designers, owner's reps, other contractors, suppliers). So, an interactive method that includes project participants is more efficient and effective. One method that achieves this end is Rummler-Brache.

That same person may take their well laid out plan and start the project. That plan will change. So, the manager must iteratively plan the project. They must revise it, some suggest daily, due to the many interruptions, delays and other changes a

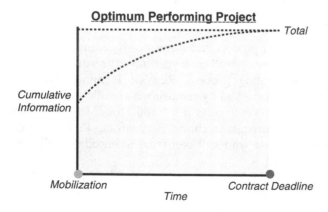

Figure 4.3 Optimum Performance Seeks "Perfect Information".

construction project experiences. In contrast, if the project reports no problems over a month, it is time to inspect the job in all aspects seriously. Construction is never a perfect process. Older managers are used to this constant review process. It is normal. They keep their project plan on their desk and not in a drawer.

As we suggested, we must define the process of planning. To start a project's planning process, we use a modified version of the Rummler-Brache methodology. (Engineers Rummler and Brache gave construction a great, if not the best, planning method.) This revised process allows us to look ahead wholly and carefully on a project. To use this method well takes some time. However, once understood, it is not hard to see its value.

A formal pre-construction planning conference takes place. This should be part of the Rummler Brache session. In a nutshell, the pre-job planning session puts together those who build with those who estimated the project. We benefit from structuring the activities with processes that work well. See Figure 4.4.

On an overall project basis, we recommend a pre-project meeting that includes:

1   Those who designed the project (such as the architect, engineer).
2   Those who fund the project (such as the developer or government).
3   Those who will use the project (such as the tenant, owner or public).
4   Those who will build the project (such as contractors and subcontractors).

This meeting is the most important one in the life of the project. A well-structured and participatory session here ensures the project will get started quickly and efficiently. Expectations, agreements and promises can be made will full information by all parties.

A project manager's monthly status report to executive management allows for solid communication between the parties and superior coaching by senior staff. Project shortfalls and windfalls are uncovered, discussed and addressed, while years of experience are offloaded to middle managers who build the work.

Weekly planners used by the site supervisors are a subset of our project planning processes. These paper planners look ahead at all parts of the project, but only a week or a few weeks at a time.

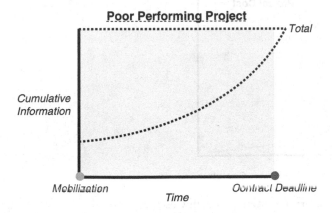

Figure 4.4 The Poor Performing Project has Slow Information Collection.

A daily crew huddle allows contractors to communicate the plan to the people who will do the work. Create a strong culture in communicating key project goals to your site personnel and you influence your means of production. Materials management (logistics) has the most significant impact on construction productivity of all issues. Studies show that this is a primary contributing factor in 70% of all wasted time in construction. Devoting regular time in this area to ensure your projects have the right material at the right time and place has a significant, positive impact. This may mean irritating time spent each day with suppliers and site people to manage the process. Instead, time is better spent in managing projects. Do not let the importance of material be lost on project managers and site supervisors – emphasise and repeat. (Figure 4.5).

There are other planning processes, to be sure. However, whatever methods you use must be repeated daily, weekly or monthly for the project to be well planned. Your first plan is not your final plan. Readjustments are typical and expected. Facing the task of re-planning every day, week or month cannot be ignored. To do so is to ask for negative surprises.

Forecasting

We do not build one project at a time as a self-performing construction contractor. Each client may expect our total devotion and resources; however, a construction firm's business model is not based on a single project or client. They build many projects, some short and some long in duration. Each has changing demands and conditions.

Reasonable forecasting addresses constraints. Here are some limitations to profitability for a contractor.

- Qualified craft journeymen and equipment operators are rare.
- Cash is limited.
- Trusted management talent is scarce.
- Equipment is expensive.

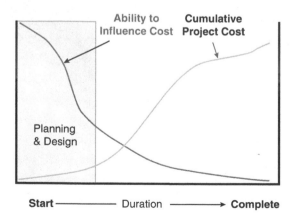

Figure 4.5 Dynamics of Pre-Construction Activities with Cost.

Source: Journal of Construction Engineering & Management.

Currently, forecasting is the missing link in construction firms. It is not a new concept that most of us overlook in our superheated race to project completion. The forecasting of demand versus supply of a company's craft workers for all projects, including the bid pipeline, helps management see problems far in advance, making them a challenge and keeping them from becoming a crisis (see Figure 4.6).

Clients do not like to hear about other projects competing for the contractor's attention. However, it is naïve not to discuss this openly. We force the discussion internally in our construction firm by forecasting demand for scarce resources (labour, equipment, supervision and cash) over our numerous projects. When we do so, we can better anticipate spikes and valleys in that demand. To not do so is sometimes to fail to keep promises and to miss project milestones.

Forecasting answers the question, *"Are the craft journeymen and equipment operators/cash/management talent available in our company for that week?"* Since construction firms do not install one job and projects have different sizes, paces and complexities, they push and pull resources unevenly. If you have an equipment-intensive business, this adds to the cost of not doing it. It is like the airline industry in storm situations. What are our priorities? Which planes should take off first? Where are the crews to fly them? Are they rested per CASA regulations? All this while the weather is clearing in spots across the country. It is complicated. The airline firms use an algorithm to determine the most efficient course of action at any moment.

Our industry does not have a ready-made algorithm for this. The demand and supply dynamic changes each week. A management discussion of the forecast on

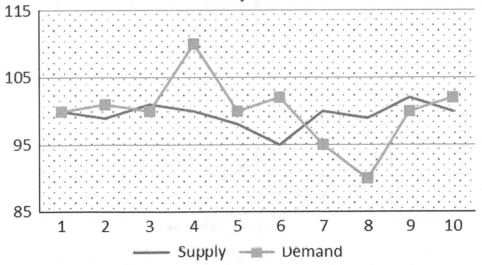

Figure 4.6 Forecasting of Demand versus Supply of a Company's Craft Workers.

Friday morning substitutes for the algorithm. We assume the planning is done correctly in this meeting, so senior management should focus on forecasting and scheduling. That is, what jobs need to be pushed forward, what projects are delayed and when to hire temporary crews to meet a spike in demand in exceptional circumstances.

Cash forecasting is another critical management activity that consistently profitable contractors undertake. They keep their borrowing limited if not non-existent. This is achieved by knowing our cash position and cash flow for many weeks into the future. Then, again, executives who see a problem can take actions now that will address it sometime in the future.

An iterative review of this report makes for proactive management of the most critical resource in any business – cash. Computer spreadsheets can be programmed to produce this look-ahead. The challenge is to estimate the future sources and uses of money accurately.

As a side note, if we continuously built the same number and size of projects, we would not need a forecasting method. This industry would be easy. An easy profession invites more competition. Let us count our blessings.

There are three parts to any forecast:

- Current demands for the resource.
- Outages for the resource (such vacations, equipment rebuilds and extraneous cash needs).
- Proposed/bid projects not yet awarded that will increase demand for the resource.

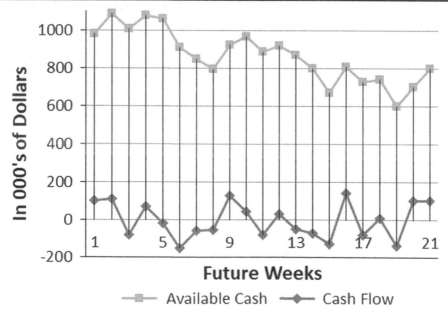

Figure 4.7 Example of a Cash Forecast. Available Cash versus Projected Cash Flow.

In the last decade, the bid/proposal pipeline has emerged as a new forecasting requirement. Clients are negotiating projects longer in time and thus deeper into the project schedule. Subsequently, contractors start projects with less lead time. Therefore, we must keep tabs on projects not yet awarded. We use a specific method to accomplish this.

In essence, effective construction contracting forecasting addresses these facts:

- Contractors build multiple projects at any one time.
- Volume is not the answer, but matching company resources consistently with client/project demands is.
- Construction firms who work steadier perform better long term than those who "hurry up and wait" or are in constant crises mode
- In practical terms, forecasting labour, cash and equipment are where most contractors should focus. Qualified labour is rare, equipment is expensive, money is limited and management talent is scarce. Thoughtful forecasting means higher utilisation and productivity for both these resources.

In our observations, forecasting is the weakest skill of PFS. This is good news in that any weakness if strengthened, is an opportunity for the most significant gain. This is an area of windfall. If you have a weak forecasting process, create and install a better one. You will see increases in time and productivity while experiencing less conflict.

Scheduling

For our purposes, scheduling is the commitment of resources to a project on a specific date. It is a promise to the client. Promises kept at a high level make for a secure construction business. Do not forget that promises kept making for better relationships if you utilise subcontractors, including future pricing.

Scheduling says, *"Since we need the resource and have it available, we are committing (promising) it to the project on a specific date."* Stated differently, scheduling results from the "what, when and how many do we need?" question. The "how many are available" is answered in the forecasting stage. The "how" is created in the planning stage.

This short-term scheduling is done at the company level for all projects. This is not a one-week schedule for a project but a six-week one for all company projects (See "time horizon" below).

Others on the project who do not plan and forecast disrupt our work schedule. They do not have the work done that is a predecessor on which we are dependent. All project shareholders must go through the planning, forecasting and scheduling act either singularly or together.

.Owners may have to insist upon the new act of joint planning, forecasting and short-horizon scheduling if they want to increase the number of projects on schedule substantially.

Planning, Forecasting and Short Horizon Scheduling should be in balance for better results. When the PFS variables are *"out of sync,"* trouble erupts. Firefighting by project executives ensues. Fire fighting in construction amounts to rework (i.e., the plan, forecast and schedule). All rework is expensive and thus unprofitable.

Put another way, if a speciality contractor is often *"out of sync,"* they will have to break promises (not adhere to project milestones and deadlines). A person or business cannot continue to break promises and expect professional trust. Lacking professional confidence in a construction contractor means one must be a "low bidder" to receive more work.

Out of Sync Examples

* *You commit to the schedule but do not have the labour available due to other projects' demands.* There is no forecasting of the scarce resource of journeymen and operators. This invites understaffed projects. Consistently understaffed projects invite rushed/overstaffed closeouts, accidents and poor client relations, among other issues.
* You have a rookie planner who misses critical requirements, so you're forecasting and scheduling process does not matter. If you do not know what you need, you do not look for it and you cannot place it on a project. A good rule:
    * The closest management level should do the planning (such as the site supervisor or project manager).
    * The plan should be reviewed and approved by a senior manager.
* This two-step process is an efficient use of the executive's time. It also challenges and coaches the inexperienced planner to be better. Managers who plan better cause less controllable problems. Thus, executives intervene less and can be more proactive.

1   *You do not include the bidding pipeline.* However, since projects have less notice to start these days (our research shows by about ½ as compared to 15 years ago), contractors must include a review of all projects bid, but not yet rewarded in their forecast. This involved a somewhat complex process of using probability and timing to accurately forecast the amount of work included in labour/cash/equipment projection.

2   *Your scheduling is done to please all clients without any thought of surprises.* That is, promises are made based on all things going perfectly. Forty hours of planned work does not equal 40 hours of executed work most weeks. Delays, mistakes and variations happen. Scheduling should be done to 90% of capacity. Construction always has surprises and emergencies. We recommend reserving a cushion of 10% for these. If your staff completes the schedule before the week is up, they will continue to work ahead (You should not expect an "early quit"). Conversely, if you are consistently missing schedule deadlines, take a look at this 90% rule.

3   *Your site and your office staff do not work well with each other.* The simplest example of this is a significant site-office conflict in your firm. Each side talks, but there is no listening and little working relationship. Sometimes, there may be a language barrier. As we are an industry of immigrants, site people may not have English as their first language. Other times, young people are intimidated by older professionals and will not ask or fully communicate about unclear areas.

There are several other out of sync examples, but these are the common ones. Let us now look at the time horizon needed for planning, forecasting and short-term scheduling.

Time Horizon

Planning, forecasting and scheduling should be done for a minimum of six weeks. Some highly respected construction firms project up to a year. This keeps the three processes "in sync." Six weeks is a starting point. Firms that are adopting a more formal (and monitorable) approach should start here. Of course, next week is highly accurate, while the sixth week from now is not. However, having that six-week schedule before you each week allows review and refinement (details filled in) on an iterative basis. High accuracy will result.

It is essential to plan and forecast continuously. This keeps the schedule realistic (promises made and kept). If we are behind on the CPM, it is still important to not be wishful when catching up. Realistic schedules reduce conflict. So realistically, add the resources needed, so you do not further disappoint your client. Long term, this is an excellent policy.

Indeed, knowing about problems days and even weeks ahead allows the issues to be managed well. A problem known well in advance is not a surprise but a challenge. On the other hand, if a problem comes to light hours in advance, it is a surprise and can induce great conflict. We believe you should understand these approaches to looking ahead, so your projects consistently come in on time and budget with less friction.

Better understanding how to put the PFS process together, communicating it, monitoring and lastly, feeding back corrective actions will allow you to catch up or take advantage of opportunities to gain project float. But, again, CPM scheduling does not plan and forecast a project; it measures.

To effectively plan, forecast and short-term scheduling, recognise that people are the only moving part and the weakest link. Remember when working with your staff that:

- An up-to-date plan, forecast and schedule only exist if written.
- Most construction people do not naturally like to plan, forecast and schedule (they would instead build something).
- A good plan, forecast and schedule are carefully put together while answering the six questions: who, what, where, when, how and why. According to quality legend Edwards Deming, the "why" may have to be repeated up to five times to get to the root of any problem.
- We must plan, forecast and schedule continuously (each week) due to the inevitable interruptions and inherent changes in construction.

Critical-path-method schedules are as standard as asphalt, concrete, conduit and switchgear in construction. Unfortunately, most of us are wary of them. However, they embody our contractual obligations and accurately show the impact of problems, so we cannot ignore them. To efficiently achieve CPM schedules while keeping critical resources profitably managed, each contractor's staff must know and follow the process of planning, forecasting and scheduling.

Short Horizon Planning and Scheduling

With 30+% of the field and office labour wasted, planning represents the greatest financial opportunity for all contractors. Therefore, reducing that average amount of unproductive time is a significant competitive edge. However, it takes a belief in planning's power.

A plan is not a plan if it isn't in writing. In the absence of a document, anyone can "paint the target." In other words, we may shoot the arrow and, wherever it lands, paint the bull's eye there. Thus, we will never build a long-term planning skill.

The field manager is the only one who can put together a weekly plan with any great detail. He is living and sleeping with the job. This supervisor knows the lay of the land, i.e., what material has been delivered, what work is ready or not, the pace of the job, where other trades are in their progress, etc.

The average foreman looks ahead about 5 hours. That applies to your competition as well as your field manager. It represents one of the great profit opportunities in construction. Looking ahead further leads to greater efficiency, less stress on everyone and better utilisation of labour and equipment.

We have been asked the question, "Which is first? Planning or Scheduling?" My answer is that the job is already scheduled per the contract deadline date. Since that is the case in all at-risk contracts, the only thing left is the planning to meet that deadline. Indeed, the owner's CPM schedule reflects a schedule and us as contractors have to plan to achieve it.

Industry studies show that an average short interval planning and scheduling effort reduces labour costs by at least 3%. Some people say it is more. We have found that it depends on two factors:

1　The consistency of written planning and scheduling.
2　The level of detail.

This written form is filled out weekly and looks ahead to the following week (semi-monthly or monthly). Since the field manager knows the project details well, it involves less than 20 minutes. Not a complex investment to make.

Almost as important, a copy of this planner should be sent to the project manager, president and other key people, so they may act on anything that is their responsibility. This action forces improved communication and coordination, not a dire consequence of this process. But, as you know, communication and coordination are never perfect in the construction industry.

The Short Horizon Planning and Scheduling System (SHPS) effectively complements construction work. It is a method specific to field supervisors and project managers responsible for managing labour, material, equipment, tools and general conditions with a limited budget and a time deadline.

If we lived in a dream world, we would have unlimited resources with no lead time. The construction business would be quite straightforward. However, we would have unlimited competition. Everyone, including the long-term unemployed, would be in it.

Since we don't live in that world, we must do some planning. Planning is critical and urgent. Two attributes that place on top of any time management list.

To further illustrate, if we worked in a gypsum board factory, would we have to do much planning? No. We would stand at a station and cut, stack and reload. This work is complicated but simple, continuous function without variation. We wouldn't need a plan "B" and each day would be the same.

The construction business has no resemblance to this. It is dynamic, changing with each day, people are different, each task is not the same, material supply is uneven and the weather is unpredictable. Planning is a must. It is difficult to get people to do it consistently and in detail. Fortunately, planning has a high payoff.

In this new century, the average foreman does not plan consistently. Thus, if your field supervisors use a look-ahead method, you will be more efficient than your average competitor.

Guidelines:

- A plan is not a plan unless it is written.
- The field supervisor has to generate a short interval plan. He knows the most about the project. The project manager, president and others review the plan for anything they are responsible for. Additionally, coaching the foreman in the finer points of planning and coordination only adds to his expertise, making him better in the long run.
- Plan your work in detail and then schedule it to follow the plan (not vice versa).
- Plan to complete 90% of your capacity. Don't overload. It leads to mistakes by pushing too hard. Rework is 4+ times as costly.
- The farther out in time you plan, the more that plan changes. It is inefficient to prepare for a dozen periods ahead. Thus, the closer in time you plan and schedule, the more accurate it is. The minimum look-ahead should be one week and it is an excellent place to start with your field managers. Many companies plan for four weeks; however, they did start with a single week SIIPS.

The best foremen do the use of a written planner and schedule. They know using this process and writing down next week's tasks is working smarter. Once it is committed to paper, it can be faxed or copied to others. From this plan, contractors have the opportunity to coach the field supervisor and make them better managers.

The field manager now has the plan filed in their field book and can refer to it often. As a result, your company will experience fewer problems and surprises on the project. This translates into better financial and schedule performance, which is the hallmark of a great contractor.

Contracting firms have to insist that this type of planning is done. Whether it is through a dictate, incentive or coaching approach, field supervisors must be convinced that this is a minimum standard to work for your construction firm. We are in a new era. The business of contracting is much too risky. Our opportunities to make substantial profits are rare. We have to take advantage of every "work smart" idea to continue to have a certain future as construction firms.

Inspiring the Ultimate Operational Mentality: Generative

This section focuses specifically on how a construction firm can manage its *people* towards a generative culture. To define, a *generative* Operational Mentality is one where all employees genuinely believe that success, no matter who gets credit, is in everyone's interest, where employees can speak up about problems without risk. All believe that good behaviours are everyone's responsibility. Every firm should be striving to create a generative culture.

When a firm has a generative culture when someone sees a problem or poor practice, they speak up and others take their voice seriously. There is no tendency to blame the messenger or cover up when there are mistakes. Instead, employees are encouraged to actively shape the company's practices by suggesting new strategies, enforcing compliance at all levels and participating in planning. Processes are developed inclusively rather than being handed down from top management. In a generative culture, success is fully internalised to the point that it is assumed to be present in every process and if it is not, someone will quickly point this out. Messengers are encouraged and supported, responsibility is genuinely shared, improvement is constantly and actively sought, problems result in inquiry and new ideas are welcomed and put into action.

In a generative mentality, all management believes that problems are their responsibility, not carelessness lower down. Employees self-enforce good practices because they want to, not because they must. Typically, the leadership of a high performing firm has some charisma and deep character. They can be trusted. Each company member likes their immediate supervisor personally and feels strongly about giving an outstanding effort. In some intensely generative cultures, we have observed that employees "look after each other."

We have found that culture makes or breaks any construction firm. This is due to the considerable influence of people in the construction contracting business. What they do decides the fate of a project's outcomes. The valued outputs by clients are safety, quality, cost and schedule. For contractors, these are critical as project outcomes determine the profitability and sustainability of that profitability over the years. Each is primarily affected by people's continuous planning, communication, execution and adjustment. Culture impacts other significant areas of a construction contracting company, such as employee turnover, overhead efficiency and problem-solving.

Project Management's First Four Practices

In construction, projects are either wins or losses. Project managers stand with others at the front line of those projects. A recent article, *What Successful Project Managers Do,* by Laufer et al. (2015), recently appeared in the MIT Sloan Management Review Magazine, offers insights about project management that are highly relevant to construction. Notably, construction was one of the industries studied. The authors developed from research on NASA's jet propulsion program and other industries, including construction, are also relevant for contractors. Their work might be used as a reframing for experienced project managers or a starting point for training assistant project managers.

NASA's jet propulsion program is similar to construction with its project-based operating model, where different workstreams have unique and unpredictable features. No two jet propulsion programs are the same and NASA's project managers pushed back against boiler-plate management practices that were too simple to accommodate the variation and unpredictability of the workflow. Building upon this experience, the researchers conducted five years of research in two dozen organisations, including construction organisations. They distilled four best practices. Effective project managers conclude:

1   Develop collaboration.
2   Integrate planning and review with learning.
3   Prevent major disruptions.
4   Maintain forward momentum.

We think that the lessons of their work should be considered by construction contractors seeking to make every project a positive one. In the following few sections, we describe each of these areas and then draw some conclusions for implementation in our industry.

Develop Collaboration

Successful projects were headed by project managers who were able to develop a collaborative, team environment. The human element is still a significant part of construction projects even as the industry evolves technology-enabled. People who work well with each other beat dysfunctional teams long-term. It is important to note that no team will ever have a propriety (patented) method in a service industry. So, those who openly share wanting the project to meet its goals help ensure its success do not hurt the organisation's competitive advantage. As each of us knows, people can be a barrier or a leverage point, depending on whom you choose.

project managers are a critical part of the formula. Successful project managers treat project setbacks as learning opportunities rather than schedule breaches. For example, they encourage team members to reveal their difficulties and struggles to help each person overcome them. However, project managers also operate inside of the organisational culture. Therefore, the culture must support collaboration.

Previously, we mentioned organisational culture; we have labelled this as a *generative-team* culture. In our experience, effective project managers will demand a collaborative environment and shield their people from pressures coming from the client. However, others will not have the skills or the courage. They need a company environment that encourages them to quietly discuss schedule problems and team failures so that the entire project team can learn and thrive.

Integrate Planning and Review With Learning

Laufer et al. describe a "rolling wave" management style used by effective project managers. This includes detailed planning for the short-term and more general, less complicated planning for the longer term. It also consists of a continuing re-evaluation of the project needs, understood through the collaboration with team members who are open and generative.

In construction, a challenge for each project manager is to capture a more significant amount of information earlier during the project life. Facilitating a discussion by all project parties is critical to get observations (objective) and opinions (subjective) into team members' minds. Then, the project manager should synthesise this information into understandable insights while assigning actions. We recommend a series of processes that provide this information at the beginning of the project and have the added benefit of starting collaboration before the project even begins. This includes, for instance, making sure that each team member is a contributor at heart and exhibits even-handed behaviour towards others.

Additionally, project managers and site supervisors who are less judgemental and just want to know the thinking behind the estimate encourage sharing by the estimator. These processes align the estimator and project staff with company goals. Establishing the practical starting point with the most current information is critical so that the project manager does not feel that they are starting on the back foot.

After a strong start, a good construction contracting firm will support a project manager by facilitating a *productive* environment. As Laufer and his co-authors point out, the "rolling wave" approach will ensure that new circumstances are incorporated, further information is integrated and new insights are considered. This will get the project up the learning curve faster, approaching 100% knowledge quicker. Moreover, it will provide a significant advantage as familiarity with the entire project makes the inevitable need for more effective troubleshooting.

Prevent Major Disruptions

Effective project managers prevent disruptions to the workflow that significantly slow productivity or increase costs. The art is to create long stretches of productive work and increase the time between major adverse events. Of course, minor disruptions will happen and experienced managers know what to do to mitigate that damage. However, they must constantly anticipate major disruptions and stay in a flexible, problem-solving mode throughout the project.

In construction, teaching project managers to identify possible sources of significant disruptions is both art and science. We often advise clients to stress essential risk management of potential adverse events. It has three components that should be part of any risk management discussion: *1) Probability 2) Severity 3) Residual Effects.* The numerical grading of those components can clarify the project team about which identifiable potential events could be significant disruptions. The highest score is the first risk addressed. Then, contingency planning may involve multiple actions once a risk eventuates. These might prompt when discussing risk management on the first day of estimating handoff and pre-construction planning.

Project managers must identify the possibility of significant disruptions. Still, it is also vital to empower them to act early by asking for advice and seeking resources or equipment to maintain workflow. A generative culture that shares potential solutions and proactive insights will assist with this empowerment.

Maintain Forward Momentum

When adverse events occur, they can have downstream effects. According to this study, it is critical to identify the event and promptly begin resolving quickly. According to the study, savvy project managers did three things consistently: *1) hands-on engagement, 2) frequent face-to-face communication and 3) frequent moving about.* These working methods allowed them to identify problems quickly and, per-haps more importantly, have a strong working rapport with the teams that would need to help them solve the issues that arise. Laufer and his colleagues cite a study of construction managers (CM) that concludes that they started working on 95% of problems within seven minutes. These CMs may not have solved all issues but starting early rallies energy and resources behind problem-solving proactively.

In our experience in construction, recriminations improve nothing. Those who reach back to past adverse events in group settings multiple times do not help the team. Planning the work and communicating that plan are forward-looking actions. Past success or failure has been determined and is unchangeable. The future is where the next win or loss will occur. So, looking forward is where the focus should be. A company with a generative culture will support project managers to maintain forward momentum by acknowledging successes and encouraging proactive management of potential disruptions, delays and disasters.

In summary, the direction of construction project management is still subject to the latest bestseller or association presentation. Sizzle sells but possibly distracts from fundamentals. This summary of specific Project Management research published by MIT, in our opinion, gets to the core of building work in construction contracting. These findings are prescriptive and fact-based. They include observations of and data about construction project managers. Contractors who align themselves to support project managers in these four areas could see a productivity increase and personnel turnover decrease.

Most construction organisations build dozens of projects a year. Since net profit before tax percentages continue to be single digit, it takes approximately four good projects to break even one bad one. As a leadership and management challenge, instilling a fundamental-based culture in a construction firm keeps everyone from making basic mistakes. These four practices are where that culture may start. Long-term, more projects should be won.

We recommend that contractors understand what successful project managers do and how organisations support these excellent frontline leaders. That is why we recommend a generative organisational culture to supplement the four practices identified here.

Workload Balancing for Project Management Teams

It is exceedingly difficult to determine if someone is productive in a service business. There is less tangible output than in a product business. Many factors cannot be measured in absolute terms. However, they can be assessed in relative terms to other similarly employed staff. Management legend Douglas McGregor's theory of X and Y workers is alive and well in the construction contracting industry. A belief in X workers, to paraphrase, is that people are inherently lazy and they must be coerced and controlled, among other things. Y worker philosophy is that people want to do well. Management should facilitate and grow that spirit. We have both kinds of employees and it may harm our staff and ultimately affect project outcomes.

Given this dichotomy, executives might make two choices, each of which produces different results. These are common construction philosophies. For example, senior leaders can choose 1) Overhead Frugality because they believe in Theory X or 2) Management Intensity of a belief in Theory Y.

Overhead Frugality

Spreading people across more jobs than less is the practice. This is a philosophy that people may need to be given more tasks to work harder. In this thought process, giving people more than they can do is supported. In our construction world, we might say if you push them, they will produce more. A practitioner might state, "I want very tough managers and I will push them to be just that". Theory X is what some construction firms follow and it may be a detriment to achieving their goals. Remember, achieving your goals means you are successful.

Management Intensity

Most people like "Theory Y" professionals. These are the people that do what is asked and rarely say no. They find work has a place in their lives. Even if they didn't have to, they would work at something. Mental engagement, along with a positive work outcome, is part of their makeup.

These are the professionals who make construction firms legendary. They work at problems until they "get it right". They lead by their work example (not by their talk). Construction firms do well or don't do well by the number of Theory Y's they have. Lose a few and excuses increase from your Theory X workers. Gain a few and things get done without much commotion. It is important to note; the client can sense the difference.

If we burn the Ys out, we will lose them. Either by health or family concerns, they will go away. Most people will slow down their pace when their bodies or loved ones tell them to. It is essential to state that parents who run families will not be continually absent. Your overworked Theory Y employee will eventually reject work as a compelling reason for absence (overtime, late at the office). No child or spouse does well alone. Additionally, when one's health declines, a career is not a top priority. Your company's work demands should not contribute to premature retirement.

Suppose we keep the possibility of burnout or negative health impacts low. In that case, Y employee intensity will be high (fewer distractions and hurdles). As a result, they will catch many project variations – windfalls and shortfalls – and thus raise productivity, quality and safety while making the job produce collectable profits. Rushing does not happen under this business management approach. Instead, the steady process will create a better long-term result, just as an accomplished craftsman's trademark.

A highly focused project manager has time to read, review and analyse daily progress. Communication is improved. Typically, they work the same extended hours, but planning, communicating and measurement are done each day. They have time to do this. In construction, this iterative process is valuable as change happens each calendar day.

Many contractors grew their business from $0 revenue. They made the company what it is today. Sometimes, this intense professional life bleeds over into their people management skills. They believe that all their people should have some of the experiences they did.

It begs the question, "Do you want the toughest project manager or one who is very profitable?" This seems silly to ask. But, on the other hand, employees are not owners and don't have an intensely focused and entrepreneurial approach; otherwise, they would be owners also. So, a rational employee management method might be helpful here.

To give a specific example of management intensity, we often hear from estimators about the endless demands and interruptions they experience. Indeed, this is frustrating for them. More importantly, they assert that their estimates would be more accurate and detailed if they had periods to concentrate on activities such as quantity take-off, costing and business approaches. Our industry has some very tough estimators. However, all company goals are achieved by allowing undisturbed and focused, making them more accurate. Unfortunately, mistakes happen in unfocused moments.

All construction firms are served well when managers understand a project and its details while having the time to act. Working in a rush leads to mistakes.

Simply put, if people can focus and concentrate on details large and small, they can drive costs lower and collect turnover quickly.

Whether in great economic times or poor ones, keeping people from overload is an essential executive duty. As we have said before, we are all in a 50-year swim. Let's not drown in the deeps or the shallows due to short term exhaustion. Drowning happens if we are overburdened. On the other hand, working in a steady, predictable way has many benefits, including better quality, less personal conflicts on the project and better health overall.

A common question in the construction industry: "How many jobs can a project manager handle at one time?" It is complicated and if it is to be answered, it needs to be accompanied by data, such as company size, average project revenue, corporate hierarchy, project length and other PM duties. Then, it would be a guess, a gut feeling and a rule of thumb. Frustratingly, your answer would be as good as anybody else's.

Before you go through all that work, we must state our opinion. First, it is the wrong question. If you answered it, that conclusion would not mean anything. Second, it is not a helpful exercise. All companies solve for cash, also known as collected profit, while safety, schedule adherence and quality (craftsmanship) are excellent. Let me explain.

To distil this down into a working measure, the question we suggest is "how much gross profit should they produce?" or "what is the financial and schedule outcome that is desired?" How about ROI? Would any of these questions cut through the fog of what a construction firm clearly wants? The question of how many projects is a blind alley. It has a thousand exceptions and is not a static answer. It would have to be changed as often as a new project was won.

Let us start with an agreed premise. Management intensity is the driver of profit dollars in construction. Therefore, if any manager or executive has time to focus, they will do a much better job catching our business's details. As you know, we are a detailed business.

A Balanced Approach Is The Goal

Too little turnover and we make no money. Too much turnover and we make mistakes including stressing our good people (Figure 4.8).

George seems to be significantly less busy than the other four PMs. Lisa is the most challenging. Part of the value of this approach is that the same data is used across the matrix and can be used for a professional and not personnel conversation about performance.

Managing the demand on managers consistently allows for better outcomes. Multiple factors have to be accounted for. They are neither severely overstretched nor under-utilised.

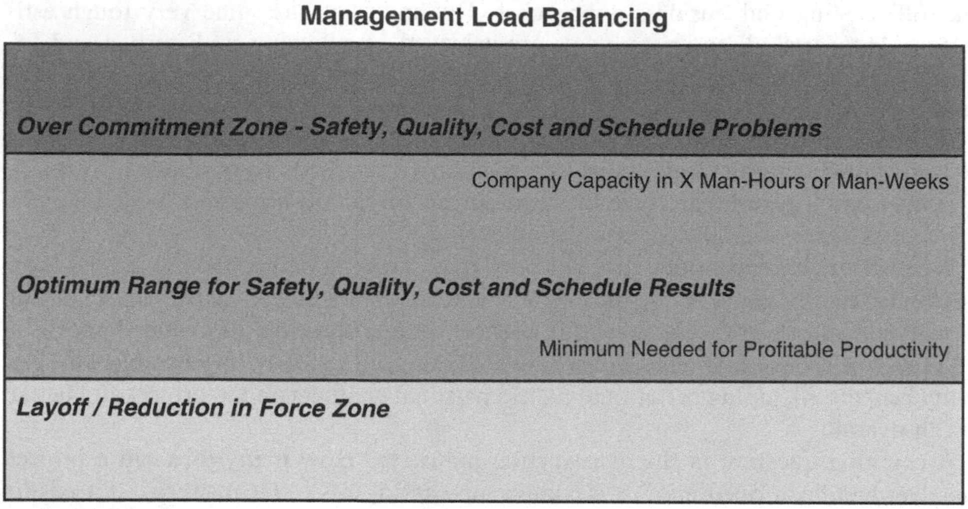

Management Load Balancing

Over Commitment Zone - Safety, Quality, Cost and Schedule Problems

Company Capacity in X Man-Hours or Man-Weeks

Optimum Range for Safety, Quality, Cost and Schedule Results

Minimum Needed for Profitable Productivity

Layoff / Reduction in Force Zone

Figure 4.8 The Balanced Load is Somewhere in the Middle.

A Steady Approach Allows for Mental Intensity

In many ventures, mental intensity is the driver of good things. For example, if an executive can raise an employee's focus and time on task, it will result in more complete and accurate work.

As a further example, sales intensity is the driver of revenue. This is critical. When most contractors' volume is down, increasing work acquisition intensity makes turnover better than average. We define sales intensity as time in front of more qualified clients. This makes for a higher probability of closing a project agreement.

Any contractor works through a slow period, so it may be an opportune time to spend extra hours fine-tuning operations, including an "executive dashboard". Once the economy picks up steam, things will be a blur again. We believe that some thought has to be given to improving your business systems in an economic downturn. To paraphrase one public official, "Don't let a crisis go to waste."

All construction firms pursue specific goals. We are an industry that has three large targets once we win a job. It is no great insight that a project manager is better served if they focus on these three. They are more controllable than the bid or proposal that the client has accepted and issued a contract based on that agreement.

The three goals of any contractor are

Productivity

Safety

Quality

Productivity encompasses cost and schedule. The typical infrastructure or building owner wants the job done – to borrow a phrase from Nike, "just do it." They want production, i.e., you to produce the contracted project with minor wear and tear on them. To constantly serve this production demand will bankrupt a contractor. Therefore, contractors have to be productive. This ensures turnover can be billed, costs are low and any reasonable schedule is met.

Safety – this is the right thing to do. Our business and project goals are not worth anyone's life or ability to work. Many adverse effects emanate from a lack of safety – morally, legally and financially.

Quality – accepted quality by the client means many things:

1   Trouble-free installations give clients a positive feeling.
2   No expensive callbacks or rework gives you a positive feeling.
3   Billings are paid and not held up.
4   Pride by a construction firm and its people.

Management intensity affects these significant outcomes. project managers who are not in a superheated hurry but are managing details with an excellent focus can keep projects graded an "A." Remember, 95% is still an "A."

Some executives strongly believe that project managers need to be overworked to be productive.

A construction company owner may feel that a P.M. should have the same experience as the owner had in building the business. They might sense that having too much to do, with less time to do it, is a management method that produces the best result.

Why the attention to management intensity? This is a good question and a fair one.

We are in a highly detailed business, probably the most complex for the net profit we earn. We have dozens of tasks to complete before the end of the job (some of us have hundreds) to complete it correctly to earn our final check. Completing one task well does not guarantee that the others will be. We have to constantly be on task with the critical path of the business or the project. Any lessening of intensity means another duty may be missed or delayed. This has a direct effect on other steps. Now, contrast our business with technology or manufacturing.

Technology: Steven Jobs produced great technology. However, how many times did he have to get it right with the iPhone? Answer: Once, he sent it overseas to be produced in significant quantities and delivered just before the holiday shopping season.

Manufacturing: Manufacturers must also get it right one time. Then, once the machine is set up, production can continue weeks on end with minor maintenance and no planning. If the machine is humming along, life is good. Remember the "Pet Rock." That creator/inventor had to get it right only once.

A good contrast to these two industries is construction. Construction shows the need for planning. Planning for technology and manufacturing firms is season to season. Months can be planned at a time. A construction firm's planning process is daily – planning is re-planning. As things are done or not done, the plan has to be adjusted. Windfalls and shortfalls have to be addressed. Material not delivered as promised happens each day in our industry. We have to compensate by a proactive and mature attitude making job progress where we can.

An average construction firm may have 150 steps to complete a project. Therefore, just because you win a job does not mean your cost will not rise unexpectedly or that you will receive payment. To put it mildly, we are in a complex industry with low margins that is human enabled. Therefore, being careful of the people management process is rational since it is one of the few controllables.

Construction people are proud and tough-minded. Asking for help or getting in touch with their stress level is not in their character. Not having enough time to do a good job is a challenge and not a problem to be shared with senior management. Taken to an extreme, sometimes a contractor won't find out about a problem until the project is 90% complete. Some leaders believe that making sure managers are running hard from job to job in a blur is a reasonable risk management strategy.

In some situations, this might evidence itself at the end of a project where profit fade, poor quality and other effects can no longer be hidden. The client may complain during the project but will not pay as expected at the end of the project. You can survive a complaint but may not survive being unpaid. If this happens on more than one job, chaos is inevitable.

Spread a manager's time and energy thinly and watch the detail orientation go away. Overhead per job will decrease but giving too much work to one (while allowing another to have fewer demands) makes for more problems than just detail orientation. The issue of uncollectable profit is one.

Severe workloads do result in problems in construction – behaviours such as rushing through meetings, pencil whipping monthly narratives or hurrying through correspondence cause oversights.

We are in the second riskiest business in the United States. Detail is where the risk resides, so detail orientation is a practical management attitude. We strongly believe in Management Load Balancing. We have made our case above. Now, let us define it. What is it and how does it work?

In a paragraph: It is the process by which we take limited management resources and evenly distribute them to keep management intensity higher on all projects. We attempt to lessen any overwhelming workload to any of our managers by distributing it equally. This will result in a consistent detail orientation; thus, projects are well built, safe while financially positive.

The better step before hiring new employees is to make sure an unbalanced load does not whipsaw all your people.

We will use the example of a project manager. This same process may be transferred to many other positions of a construction firm.

Significantly, we cannot change weekly who manages or is involved in projects. Once a person is assigned to a project, keeping the same person managing it until the completion can be the only rational approach. We want to make sure that we do what we can to foresee any significant overwhelming events.

Also, we certainly want to know when someone may have extra time on their calendar. This under-utilisation can happen while the employee says nothing.

To properly execute management load balancing, it is essential to remember it is a measurement that we are after. This coupled with our subjective feel for our people, is needed to assign projects with confidence.

As an example, if one project manager will be working with three new clients (ones he or the company has never worked with before), we might be wise to assign one of these customers to another project manager. It would lessen the steep learning curve of client relations and work styles. Better customer satisfaction and more profitable projects would result.

Process

Again, there is no industry standard here. As all construction firms are unique in some ways, we are better served by knowing our historical norms. To that end, track the last three years of management metrics a quarter at a time. A few possible measures are:

Of New Clients.

Gross Profit per Person-Hour (hourly staff).

Gross Profit per Person-Week (salaried staff).

Field Person-Hours per Person-Week.

This is just a sample list. By trending these and finding the normal range for your firm, you will know what acceptable performance is in numeric form. Once implemented, you can easily extend it for your organisation, its different departments and individual professionals.

Using multiple factors to determine who is overburdened and who is not allows contractors to keep all personnel working steadier (Table 4.1).

Take the above example and answer this question, "Who should manage the next project?" Lisa and Jerry would not be my first choice. They have the most demands on their time. Although, we are sure they would say "yes" due to their Theory Y nature.

Executives who keep people from being overwhelmed or, conversely, others from taking a rest on company time are leading their firms to a steady focus on safety, quality, cost and schedule.

We would look at Nancy and George to manage the next project. They have the capacity in several ways. Indeed, any serious discussion about why would involve this chart. It is rational to start with the numbers in making management decisions. However, executives have to include their instincts and gut feelings when making final determinations.

This starts a discussion but does not answer the question of who should manage this upcoming job. Typically, other factors exist that direct the decision to a more thoughtful conclusion. However, starting from a place of reality makes for a more effective and efficient discussion.

If Nancy or George is a Theory X employee, they may argue against managing this latest project. That is their nature. They may cite their overwhelming workload (not actually from a quantitative basis) and other demands on their time. On the other hand, a week or sympathetic executive may take the easy way out and give a "Y" project manager the job. In this way, the X's abuse the Y's.

Sometimes, managers will act busy and tell you how busy and overwhelmed they are. If you don't make management load balancing part of your executive process, you may believe them. This quantitative approach does not lie.

Again, losing the Theory Y managers has a more significant negative effect than losing a Theory X one.

You may want to take this kind of thinking further. This exercise has to make financial sense. Without collected profit, a construction firm is either troubled or extinct. To that end, creating a profit and loss statement per project manager to confirm a targeted return on investment may be advisable.

Table 4.1 Example Management Load Balancing Tracking Form

Name	Projects	Active Projects	# Of New Clients	Total Under Contract	Hours/Weekly Meetings	Unearned Revenue	Total Distance to Projects	Unspent Labour Hours	Largest Project
Bill	9	4	1	$7,522,641	12	$5,602,112	500	$30,585	$4,878,038
Jerry	7	5	1	$10,816,653	15	$7,500,292	430	$23,035	$3,003,912
Lisa	4	7	2	$11,625,682	10	$8,152,252	390	$18,945	$4,336,287
Nancy	3	4	1	$5,231,355	7	$3,847,827	300	$6,781	$1,471,385
George	2	6	0	$3,022,010	4	$2,001,280	90	$2,097	$1,939,540

Keeping people from exceeding the normal range and you have done all you can to keep people from being burned out, overwhelmed or just too busy to pay attention to the details of the complicated challenge of building work. All construction firms benefit from taking care of each element or step in a timely and complete manner.

With the advent and growth of Design-Build, construction firms are managing more personnel and processes. For these firms, balancing the load between all managers is more important today. In one way, management load balancing is planning your project team's work over the medium and long term; it is a controllable activity. You either do it, or you don't clearly and thoughtfully. For some of your employees, it may be a sign that you genuinely care about them.

An overwhelmed employee is a constraint to your business process. The Theory of Constraints states that identifying and then removing a bottleneck will speed up the process. You are only as fast as your most restrictive restraint. Technical and business operations, if sped up, gain a competitive advantage for the contracting firm. They convert into a better ROI and a cleaner balance sheet and profit/loss statement.

The last letter wins in most legal disputes, a preponderance of evidence notwithstanding. The construction person who has time to follow up with a pointed argument has a significant impact in deciding a legal disagreement. But, again, more extemporaneously created documentation can be powerful evidence also. Both of these are an outcome of less distraction and more focus on a project.

An overwhelming month of work can be deadly to a construction firm and its prospects. Opportunities and threats to your company may be missed.

The manager who has a focus on:

1   Producing acceptable work for the client,
2   Observing and notifying clients of changes to the contract scope and
3   Capturing delays by others and enforcing our contractual rights.

They will deliver more earned turnover and cost-reducing project performance.

Whatever your view of people and their love of work, managing your firm's workload is an effective way to keep people focused. The construction industry's "too hot, too cold" problem can cause a lack of focus. Controlling the controllables is what executive management is all about. Take a long look at management load balancing as a formal solution to keeping your people from being overwhelmed at times while others take a "mini-vacation."

Better and consistent work outcomes will be the result. In our complicated industry, this can only benefit any construction firm, whatever its long-term goals are.

Using DISC to Select Better Construction Crafts Persons

To understand people's personalities, we are best served by using a proven system of analysis. There are several ones on the market and available to contractors. However, none is more straightforward, tenured or defendable than the DISC system. The DISC Profile Methodology is a helpful device to understand an individual's behavioural pattern and what emotions they might feel. This system provides labels, definitions and a structure to explain different visible behaviours and internal feelings. This aids contractors so an employee's personality strengths can be directed and weaknesses can be augmented for everyone, including the employee. Equally as valuable when analysing a prospective employee. Many contractors have experienced "we hired him from his resume but fired him based on his behaviour!" DISC helps project the future employee's behaviour in the job. In other words, DISC helps explain the mysteries of people.

Several caveats should be kept in mind when analysing DISC information. First, no one should attempt to change their personality profile due to DISC information. To do so would probably be a waste of effort. It might work for a short time; however, the strain would eventually force you to return to what is natural. However, people undergo rapid personality changes in what psychologists call psychological trauma, such as an event of excellent luck or, conversely, misfortune. For example, winning the lottery, the death of a loved one or a divorce.

Second, there are neither "superior" nor "poor" profiles. We have found successful, incompetent, happy, miserable, personable and unfriendly people with every one of the DISC behaviour profiles. An ingredient that frequently appears in people who are successful in their jobs is all personality types, but the person's awareness and accommodation of the strengths and weaknesses of their personality, whatever the style. To say it in two words, self-awareness. The DISC process aids in self-awareness and can recommend tactics to accommodate one's personality to win at their job more often. Specific behavioural patterns adapt more quickly to certain types of employment; the DISC system is beneficial as one source of information for making staffing decisions. However, be careful; all behavioural patterns have proven successful at work.

Third, the information contained here is general and therefore, should not be used as the sole criterion for making personnel decisions. The DISC does not measure I.Q., experience or talent. That determination is made by checking references, administering intelligence tests and trusting your gut feel.

The DISC Profile Methodology is based on the work of behavioural scientists Marston and Grier. Marston wrote a breakthrough book named "The Emotions of Normal People in the 1930s. He asserted that personality has four dimensions:

D – Dominant or Driven Personality
I – Influencing or Inspirational to Others
S – Steady or Systematic about Work
C – Compliant or Conscientious of the Rules

Each individual uses each of these dimensions to relative degrees, which may or may not vary depending on the state under which the individual is functioning. For example, the behavioural conditions, which the DISC evaluates, are perceived expectations of others (and job), pressure and perception of self. Most individuals will use one of the four dimensions more than the others under a state of stress. This dimension is called the "high" or primary characteristic for this explanation.

Dominant or driven individuals attempt to shape the environment by overcoming opposition to achieve results. These persons may be quick decision-makers, good problem-solvers, unafraid of uncertainty and new challenges and adept at getting results. But, on the other hand, they may also be impatient, pushy, easily bored, quick to anger and uninterested in detail.

Highly driven persons usually prefer to work in situations with authority, challenge, prestige, lack of supervision, various tasks and opportunities for advancement. They often are effective when working with others who are more inclined to calculate risks, more cautious and more patient with others.

To be more effective, highly dominant individuals should seek complex assignments, which require techniques based on practical experience, make an effort to understand that they need cooperation from others to succeed, identify reasons for their conclusions and pace themselves. Occasionally, they need a shock if it becomes necessary to make them aware of the need for different behaviour.

Highly influencing or inspirational individuals attempt to shape the environment by obtaining the alliance of others to accomplish results. These persons put effort into making a favourable impression, helping and motivating others, entertaining people, generating enthusiasm and making personal contacts.

These inspirational individuals usually prefer to work in situations where they have social recognition, group activities, freedom from controls and details, freedom of expression and favourable working conditions. They are often most effective when working with others who speak out candidly, research facts and details, concentrate on follow-through and deal with things or systems rather than people.

To be more effective, highly influencing individuals should manage their time if their dominance or steadiness dimensions are low. For example, they should attempt to consistently meet deadlines, apprise others on more than people skills, be more objective and be firmer with others if the influencer's dominance dimension is low.

Highly steady or systematic individuals cooperate with others to carry out tasks. These persons are usually patient, loyal, good listeners and slow to anger. They may dislike noisy arguments and will calm down excited people. They are capable of performing routine or specialised work.

High systematic individuals usually prefer to work in situations with security, little infringement on personal style, infrequent changes, identification with a group and sincere appreciation for their contribution. Therefore, they are often most effective when working with others who delegate, react quickly to unexpected events, work on more than one thing, apply pressure and flexible work procedures.

To be more effective, these steady individuals should condition themselves for change. However, before it occurs, they should seek information on how their efforts contribute to the overall organisation, seek co-workers with confidence, ask for guidelines to accomplish tasks and seek out those who encourage creativity.

Highly compliant or conscientious individuals attempt to promote quality in products or services under whatever circumstances exist. These persons usually follow directives and standards, concentrate on details, comply with authority, criticise and check for accuracy and behave diplomatically or formally with others.

Highly conscientious individuals usually prefer to work in situations where others call attention to the high compliance individual's accomplishments, security and standard procedures, identification with a group and the absence of frequent or abrupt changes that might compromise quality. Thus, they are often most effective when working with others who delegate, seek more authority, make quick decisions, compromise and use policies as guidelines instead of rigid rules.

To be more effective, high compliance individuals should seek out precision work, tolerate some conflict, appraise people on more than their accomplishments, ask for guidelines and objectives and set deadlines for planning and deliberating.

As a rule of thumb, most people possess two personality traits above the "mid-line" or average emotion. They have two below. If three are above or below, this may indicate a change of circumstance – good or bad. On the other hand, if all four personality traits are close to the mid-line, then the same is true again.

DISC does not discriminate against people due to their creed, colour, race, national origin. In the parlance of Human Resource Management, it meets the 4/5th's rule. Valuable in today's employment climate.

A recent study was completed by some vendors of a DISC instrument process. The focus was what makes the best labour or craftsperson? Some analysis was done of hundreds of individuals who work with hands in hourly positions. The industry studied was not construction, but the results do point to insights that cannot be ignored.

The study assumed that the best employees have the following attributes:

1   **Safe** – safety conscious and have an above-average safety record.
2   **Productive** – Do not waste or make many mistakes in their work.
3   **Loyal** – Are loyal and have been with their employer several years.

This sounds like an above-average construction craftsperson or labourer. Personnel records and other employment information were collected. A review of the best employees was completed.

The results state that two DISC emotions correlate to the three attributes above.

a   A low "D" – this person seeks much information. They do not fly off in rage but take a careful assessment of the situation. Hence, they do not quit abruptly. Instead, they take time to make decisions.
b   A high "S" – this person is friendly and process-oriented. They do work methodically, therefore, less waste and mistakes. But, again, being nice lends itself to forgiving people, including you when a supervisor might say or do something offensive.

The extremes of any DISC rating scale from 1 to 28 are not valuable in a workplace environment. Therefore, someone with a 1 to 4 or a 25 to 28 rating should be given more scrutiny as to their fit into your office or site.

If you choose people to hire in a labour or craft position, DISC will help you make an informed choice for the long-term benefit of your company.

In summary, the DISC methodology can join other tools and techniques to make managers work better with people. The classifying system can be an essential aid for:

1   Increasing awareness of others and yourself.
2   Selecting people to work together.
3   Conflict Management.

A DISC assessment takes about 10 minutes to complete and a half-day to understand the system and process. Its value is that it saves the contractor years of trial-and-error working with people. Therefore, we commissioned a recent study to research, analyse and conclude the behavioural profile of excellent site staff and sub-contractors.

It was assumed that the ultimate value in the construction industry is quality and timely work. Superior craft persons and operators produce that with safety. Without crews of this ability, projects suffer.

Consumers and Developers alike do recognise that value and will pay a premium for these types of professionals. Therefore, they will contract with companies with more craft skills and safe practices over the low bidder.

We researched and queried the industry about measurable factors that typify an above-average craft person and operator. After long, iterative discussions, our firm concluded the following:

- Safe – no lost time accidents in last five years or 10,000 hours worked.
- Efficient – top 25% in consistent efficiency and minimal rework.
- Loyal – they have been continuously employed with the company for a minimum of five years or 10,000 hours.

Subsequently, we contacted companies from a target list. They are 1) active and inactive clients 2) newsletter patrons and 3) other interesting contracting firms.

After receiving responses from those interested, we qualified companies who possessed:

1   A full-time human resources person. This criterion was in place so, that the control of the study was kept high.
2   Non-supervisory labour (craftsperson and operator) is a minimum population of 50 in the company.

We asked them to submit five people in the top 25% cumulative of the three areas outlined above in each qualified firm. This count of five people kept the statistical significance of any one company modest.

We used a standard DISC assessment profile from a well-accepted provider. In addition, each craft person and operator completed a profiling instrument.

Results:

High correlation to a low **"D"** dimension. – lower than the midline.

High correlation to a high **"S"** dimension. – higher than the midline.

Low correlation to a consistent **"I"** dimension.

Low correlation to a consistent **"C"** dimension.

Interpretation

It is our practical experience that the craft person and operator who is concerned about safety and efficiency and who possesses a sense of loyalty would have to have this profile (Low "D" and High "S"). Its other words, it makes sense.

A previous general study concerning labour in the transportation business also concluded this general theme.

As a practical, day to day matter, these site professionals:

1   Are easy to manage if you are consistently organised in planning and scheduling.
2   Are less interested in going into business themselves and competing against you.
3   Seek more information before making a decision. Do not make snap decisions that will have later negative consequences
4   Are more organised in their approach and rarely miss a step (rework).

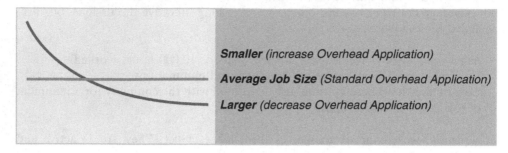

Figure 4.9 Overhead cost application is disproportionate between small and large projects.

The Case for a Small Projects Division

Small projects represent an opportunity for all construction contractors. The largest built environment firms may be dismissed as too troublesome, but we believe they should be reconsidered as a strategic choice.

We have worked with several firms, primarily large firms, that have adopted this direction. The reasons that they cite are compelling:

Small projects never go away: Recession, stagflation or boom, small projects are present in all economies. Additionally, small contracts may make the difference between a good and poor year when project activity in other states and contractors travel to yours to compete.

Small projects are an intense training ground. Young people and those entering construction after a stint in another industry may need low risk but hands-on experience. This division offers that.

Small projects can keep an older valuable manager challenged. This new venture could avoid not losing a valuable manager to another firm due to boredom or lack of challenge. Certainly, those more senior employees who love to mentor but want a small scope are good seconds in command.

Small projects are mostly no retention and high-profit margin contracts. The financial beauty of small projects is that their terms and conditions are less onerous. At the same time, the capability of a more prominent firm backs them can add to double-digit margins that small contracts typically produce (Figure 4.9).

Small projects need to be built by a low overhead division. See Figure 4.7. As shown in chapter 3, overhead sizing of small projects can be up to three times the average amount of cost to the manager but about 85% on large projects. If you feel a small project focus has strategic advantages, then a small division led by a tenured and talented manager can win.

5 Financial Management

Finance is the life's blood of a construction organisation; without, the firm dies. So, it is critical to analyse and execute accurate and sensitive algorithms and equations to protect against negative trends. Some of these equations may have first been developed in other industries. However, with its unique dynamics, the construction industry cannot be financially governed the same as other industries. Therefore, we assert that many governing principles do not apply to construction contracts. See Figure 5.1. For example, see our section below on cost sensitivity, i.e., debt leverage is a risky strategy.

Finance supports the rest of the construction organisation. The coordination of sources and uses of cash determines the options a contractor has. Most importantly, project operations generate 100% of the turnover and approximately 90% of the company cost. This should be the primary focus and a significant responsibility of a project manager. This chapter will present and apply several basic financial approaches that many construction firms utilise. A well-organised financial management process collects, computes and then disperses financial information accurately and efficiently. Knowing where you stand financially is "flying with radar". However, not knowing can be deadly. The more our financial decisions are based on accurate financial tracking, the better our choices. This chapter is a starting point for your understanding and improvement efforts.

The Basic Financial Vocabulary for Constructors

Since not all financial approaches apply to construction contracting, what are the ones that do? Let us start with 20 basic measures and ratios. These will help with your understanding.

Current Ratio – The current ratio is probably the most frequently used financial measure for the contractor. This ratio indicates the number of times existing assets cover current liabilities.

Generally, banks and bonding companies use a 2.0 to 1 current ratio as the minimum standard. However, most believe that this standard is not always applicable to contracting since contractors may exploit more significant operating opportunities by minimising the existing assets by the nature of the business. Contractors usually have a relatively small proportion of their total assets invested in inventory compared to the relatively high ratio in other industries.

A current ratio of 1.5 to 1 is entirely adequate for most contractors. Thus, a minimum of 1.5 to 1 is recommended. With a current ratio of less than 1.5 to 1, the contractor may have difficulty meeting current obligations due to insufficient working capital.

DOI: 10.1201/9781003290643-5

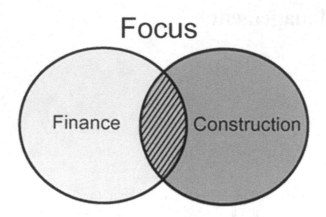

Figure 5.1 All generally accepted financial management practices and metrics are not 100% applicable in the Construction Industry.

On the other hand, a contractor with a current ratio consistently higher than 2.0 to 1 indicates outstanding financial operating strength, but the company may become inefficient and stagnant. When the current ratio approaches or exceeds 2.5 to 1, the company is very liquid. In such cases, the excess current assets should be invested in other profit-generating ventures.

Quick Ratio or "Acid Test" – The quick ratio, or acid test, measures the ability of a firm to cover its current liabilities by using its cash and accounts receivable without converting inventory or other current assets to cash.

Most financial analysts agree that the ratio should be in the range of 1.0 to 1 and 1.5 to 1. Contracting firms with a quick ratio of less than 1.0 to 1 will not usually be able to meet current obligations without converting inventory or other existing assets into cash or borrowing. On the other hand, if the quick ratio is more than 1.5 to 1, the firm should be too liquid and invest excess liquidity in other profit-producing ventures or maximise cash turnover.

Current Assets to Total Assets – The relative liquidity of a company's total assets is essential to most financial analysts.

The ratio used to measure this relative liquidity is the current assets to total assets ratio. Thus, for example, a ratio of less than 60% may indicate an excessively high investment in fixed assets, whereas an amount higher than 80% suggests that management has decided to minimise investments in equipment, vehicles, real estate or other fixed assets.

The recommended range for current assets to total assets is between 60% and 80% for those contractors who own (rather than lease) most of their fixed assets. Equipment-intensive contractors will understandably have a ratio towards the lower end of the range.

Working Capital – Working capital is another measure of the relationship between the current assets and the current liabilities.

Working capital is determined by subtracting current liabilities from current assets. It measures the funds available for future operations since it is the amount left after meeting current obligations. Working capital may be increased, if necessary, in three ways:

1 Generate and retain operating profits.
2 Borrow money long-term.
3 Selling owned assets and leasing them back to the firm.

Working capital is subject to certain distortions, and misinterpretations since some current liabilities which reduce the working capital number can represent cash (someone else's) available for operations. However, the firm's working capital must be sufficient to finance current sales and operations for long-term health and growth.

Net Excess of Billings or Cost – When negative, this figure shows that the company has not actually earned all of the costs associated with its billing. Such a circumstance is desirable because it indicates that clients are financing a portion of the contractor's work. However, since the amount of billing above costs is carried as a liability, a net negative position may have the inconsistent effect of lowering the current ratio, quick ratio and working capital when the company has sufficient cash for operations. Such a condition is all right so long as the company's backlog stays constant or increases; however, a decline in backlog can cause a severe cash shortage even though the financial liquidity indicators may improve.

When the net excess billings or costs value is positive, which is more common than negative, the company is financing the majority of its workload.

Although the amount of costs over billings is carried on the books as an asset, the contractor may have difficulty producing the cash to cover labour and material expenses incurred but not billed. Contractors operate successfully with either positive or negative net excess billings or cost figures, but each has its advantages, requirements and pitfalls which must be understood and managed. Therefore, changes and trends in this figure should be monitored carefully.

Age Of Costs & Estimated Earnings above Billings – This measure shows the number of days the company has invested in job costs that have not been billed. Comparisons to the age of billings above costs and estimated earnings and other indicators are helpful to assess the company's balance of expenses and billings.

Age Of Billings above Costs & Estimated Earnings – Billings above costs and estimated earnings is a current liability that requires that the company provide the services in the future which have already been billed. Therefore, comparison with the age of costs and estimated earnings over billings and other indicators are useful to assess the company's balance of costs and billings.

Age Of Trade Accounts Receivable – This measure indicates the number of days on average that the company has funds tied up in trade accounts receivable. The challenge is to lessen the time and thus, increase cash flow.

Age Of All Receivables – Retainage (or retention) and notes receivable are not as easy to accelerate for collection as are accounts receivable. Nonetheless, it is helpful to know the number of days the company has invested funds in all receivables.

A standard range is not feasible, but the trend of the age of all receivables should be monitored to determine if the amount of cash tied up in receivables is increasing or decreasing and, if so, determine the probable cause.

If the figure exceeds 50 days, the company loses the opportunity to invest funds more productively. If the goal is to increase available cash, decrease this number.

Age Of Material Inventory – The age of material inventory indicates the number of days that the company has cash invested in inventory. As a general rule, contractors who have significant material inventory should range from 30 to 45 days.

Age of Trade Accounts Payable – This figure indicates the number of days between a bill's time and when it is paid. To take advantage of trade discounts and to remain in the good graces with those who extend credit to the company, it is recommended that the age of trade accounts payable not exceed 30 days.

Total Debt to Net Worth – The total debt-to-net worth ratio, often called the "debt-to-equity" ratio, is an essential measure of the owner equity's relative amount. The maximum acceptable debt-to-equity ratio for most contractors is in the range of 2.5 to 1 and 2.0 to 1. Creditors should not have more than twice as much invested in the company as the owners.

A ratio of less than 1.0 to 1 indicates a strong equity position, but management may be missing the opportunity to use financial leverage. The absence of this monetary efficiency could imply a lack of incentive for future growth, although the potential is present.

Times Interest Earned – The times-interest-earned ratio is a "coverage ratio". It shows how often earnings cover interest charges and how much earnings can fall before endangering payments to creditors.

Otherwise, the distortion would result from omitting one interest coverage multiple and ignoring the tax-deductible status in interest. Therefore, the ratio is calculated by taking the profit before tax (operating income) plus interest and tax and dividing it by interest expense. The numerator of this calculation includes funds used to pay interest charges and taxes.

Total Overhead to Turnover – This ratio indicates total overhead's effect on profitability since the operating profit margin equals the gross profit margin less the total overhead to revenue. Also, the total overhead to turnover ratio is a convenient indicator to watch for trends. An increasing overhead ratio indicates that sales volume is falling relative overhead costs are increasing, or both. A decreasing percentage may accompany the ordinary disruption as revenues expand, straining the overhead capacity.

Total Overhead to Total Direct Cost – The overhead to total direct cost ratio is an important indicator. The measure shows that total direct costs must be marked up to recover the entire company overhead. Most contractors assume that the overhead should mark up each job to cost ratio to recover overhead. Unfortunately, overhead recovery is not simple but involves material ratios to labour, job size and other factors. However, trend analysis of total overhead to total direct cost can reveal changes that should be incorporated into a company's tendering policy.

Return on Working Capital Before Taxes – Return on working capital is a valuable indicator of working capital turnover and other measures to determine how effectively management uses working capital and how productively. We recommend a range of 40% to 60% return on working capital.

Return on Assets Before Taxes – This measure shows how management uses and manages all the assets to make money. The categories contributing to return on assets are accounts receivable, material inventory and fixed assets. As with some other indicators, differences among companies make a common value difficult; however, several analytical techniques are beneficial.

Return on assets (ROA) is approximately the product of asset turnover times the operating profit margin. Thus, management may consider ROA as an indicator of a combined asset (asset turnover) and productivity (operating profit) metric. ROA can be improved by divesting non-earning assets or more efficient employment of the assets.

Return on Net Worth Before Taxes – In determining which investment out of several, an investor is primarily concerned with the likelihood of a return commensurate with the risk associated. Thus, the investor will expect the opportunity for a higher return with more risk.

The owner of a contracting firm is an investor in a high-risk venture and should expect a high return. Given the inherent risk in the construction industry, we recommend a pre-tax return on equity (or return on investment) range of 20% to 40%. A lower return cannot justify the subjection of the owner's investment to the risk of loss.

Gross Profit Margin – The gross profit margin is a composite indicator of the effectiveness of pricing and productivity. It reflects management's ability to price both competitively and profitably in the company's markets. The gross profit margin also reflects management's ability to manage direct costs and work productively.

Gross profit margin should ideally be analysed for trends since various factors, such as accounting methods, can affect it. For example, an increasing gross profit margin may indicate an expanding market with the promise of increased competition, or it may indicate effective use of job costs reduction efforts. Conversely, a decreasing gross profit margin means a declining market, intense competition, or inefficient job management.

Operating Profit Margin – The operating profit margin is one of the critical measures of management effectiveness. If appropriately computed, the operating profit margin shows how well the company performs its core business, unaffected by tax consequences which may fluctuate with the dollar amount of profit.

The average operating profit margin in construction is too low, given the risks of the construction industry. Many factors affect an individual contractor's operating profit margin, so a standard value is not feasible. However, a likely result is that an operating profit margin (including a fair owner's compensation scheme) is in the mid-single digits.

Net Profit After Taxes – The net profit margin shows the company's bottom line profit performance. Net profit means profit after taxes, so the dollar value of net profit is the amount available for reinvestment in the business as retained earnings. Many factors affect the net profit margin from one company to another, so a standard measure is not feasible. However, the net profit trend should be monitored carefully and management should constantly interpret the reasons behind its movement.

Working Capital Turnover – A financial measure of considerable significance to contractors is the ratio of net revenues to working capital or working capital turnover. Working capital is the difference between current assets and current liabilities and its turnover indicates how efficiently the company is using its working capital.

For most profitable contractors, working capital turnover is between eight and twelve times per year. A working capital turnover figure much lower than eight may indicate that the company is not using its operating resources vigorously enough in pursuing jobs at reasonable tendering prices or in executing jobs won. Conversely, a figure higher than twelve may indicate that the company is taking on more work than it can handle efficiently. Also, increased working capital turnover often accompanies underpriced jobs that add to a higher but less profitable sales volume.

Asset Turnover – This measure is one indicator of management's efficiency in the direction of resources. A standard figure cannot be given for asset turnover since too many variables influence it from company to company. Scrutiny of trends and changes in asset turnover are, therefore, more advantageous.

A declining trend in asset turnover indicates either a decline in revenues or an increase in total assets. Management should consider a reduction carefully to determine whether pricing is too low, tendering efforts are slipping, or investments are being amassed in a declining market. An upward trend should be evaluated to determine if the company's asset base is adequate to keep up with increasing sales.

Constructors Should be Cost-Sensitive

Since the net profitability percentage of the average construction firm is a single digit, it is appropriate to consider cost as a primary financial focus. Therefore, one of the critical understandings of constructors is to know Loss/Turnover Replacement. See Table 5.1.

Each loss in construction has to be paid for from net profit. This illustrates how an average construction company must generate substantial turnover to pay for mistakes and theft. These losses are easily identified, such as safety, job cost overruns, larceny and rework. Other types of losses are less clear, such as inefficiency, poor morale, lack of paperwork, unrecognised change orders, unrequited job site favours. This illustration effectively portrays the considerable cost to companies in an effort to expend to make up for the loss. If an interest charge for debt is being paid, it has the same loss/replacement ratio problem.

Debt Cost and Risk

Many realise that all money has a cost. In conducting your business, your funding costs will fall into two major categories:

> The cost of borrowing: Fees or interest charged by the bank and paid by you are apparent. There are agreements in writing and document trails of transactions – bills and payments – that are easy to understand and trace.

> The cost of opportunity: Not having funds to invest is a lost opportunity. $0 invested means $0 returns received. The cost of living, including the cost of retirement, remains.

However, in our low margin business, there is less opportunity than in other industries. Contractors do not become handsomely rewarded consistently by being first movers. There are too many uncontrollable circumstances in the industry.

This brings us to the first rule of debt: Always assume that income will be interrupted or reduced. Before COVID, some people erroneously thought their income would grow

Table 5.1 Loss/Turnover Replacement

Mistake/Theft	Profit needed to pay for Mistake/Theft	Turnover needed to pay for Mistake/Theft Company makes 4% Net Profit
$100.00	$100.00	$2,500
$1,000.00	$1,000.00	$25,000
$10,000.00	$10,000.00	$250,000
$100,000.00	$100,000.00	$2,500,000
$1,000,000.00	$1,000,000.00	$25,000,000

and took on more debt. Things have become difficult for people with high amounts of debt. A poverty-born contractor created this rule, and it never fails those who need it.

During any economic recession, significant investment opportunities decline. If you understand that all money has a cost, you will pay off all your debts and know that you have saved the future cost of finance charges on those debts. You might even pay future bills if you can receive a discount from the other party for early payment. This is the best use of money in difficult times. This reduces the cost of business. The savings is equal to a proportionate gain on any investment.

Conversely, if you are a person who has a large amount of cash, you may be able to purchase valuable but distressed assets at a discount. This often occurs in every economic recession. It is a leverage point.

In business, these principles are just as crucial as they are in personal finance. Keeping debt out of a construction business is a long-term springboard to a better business. There is no other approach in a single percentage point and high-risk enterprise. Anyone can grow volume with loans in construction, but there will be little profit to pay loans back.

Credit lines can be a crutch for many poor business practices. Try to keep yours at zero for most of the year. Doing so will keep your management discipline high. Over the years, continuously using the line should tell you something about your customers, type of work, areas you work in, or even your financial habits. Consistent use of a credit line is an indicator of something negative. In the construction industry, it is the leading indicator of portending bankruptcy.

Project Return on Investment: A First Principle

What is the single metric to measure financial success on a construction project? As you know, margins have historically been in the single digits. There are numerous factors for this, not the least of which are risk factors and a high number of competitors.

How does one compete and stay financially viable? These are good questions for all construction contractors. People have different answers, as is reflected in the bid results on any given day. Some contractors are satisfied with making a living, while others are trying to build wealth. To be sure, to compete against those who just want to make a living, you will have to be satisfied with just that. Their bids have little profit margin; therefore, there must be less in yours.

Our answer: use Project Return on Investment (ROI) as your first and last measure of profitability. In this updated article, we will explore its characteristics, calculation and application to construction organisations.

Is not this just ROI? Yes, it is, but we feel it is critical to add the annual concept. Projects' durations are short, long, or in-between. Illustrate its power; see the following example in our books. A 5% project margin can translate into 12%, 45% or 135% PROI depending on the timing and number of financial transactions. These vastly different outcomes illustrate the significance of negotiating better terms and conditions. See Table 5.2 and Table 5.3.

What is the Proper PROI for Construction Contractors?

To establish a proper PROI, we must first agree that construction contracting is a high risk. The risk is evident in bankruptcy statistics. Construction contracting is still

Table 5.2 Project Financial Inputs with Terms and Conditions

		Job 1	Job 2	Job 3
Contract or bid amount of job		$484,370	$523,430	$393,800
Costs:				
Material		$228,000	$289,000	$225,000
Labor		$186,000	$161,000	$114,000
Overhead on Materials/Subcontracts	rate = 8%	$18,240	$23,120	$18,000
Overhead on Labor/Equipment	rate = 15%	$27,900	$24,150	$17,100
Total Costs		$460,140	$497,270	$374,100
Profit expected in dollars		$24,230	$26,160	$19,700
Profit expected as a percentage of contract		5.00%	5.00%	5.00%
Length of job (in months)		3	4	2
Expected lag in receipt of funds after billing customer (in months)		3	1	2
Supplier (for material) credit terms (in months)		1	1	2
Current working capital (in dollars)	$300,000			

Table 5.3 Project Return on Investment Results from Table 5.2

	Job 1	Job 2	Job 3
Length of job (in days)	90	120	60
Length of receivables commitment (in days)	90	30	60
Length of payables (in days)	30	30	60
Net material cost days	60	0	0
Net labor cost days	90	30	60
Net investment rate per day	$5,112.67	$4,143.92	$6,235.00
Material costs per day	$2,736.00	$2,601.00	$4,050.00
Labor costs per day	$2,376.67	$1,542.92	$2,185.00
	Investment in Each Job		
Total investment in material	$164,160.00	$–	$–
Total investment in labor	$213,900.00	$46,287.50	$131,100.00
Total working capital investment	$378,060.00	$46,287.50	$131,100.00
Profit generated	$24,230.00	$26,160.00	$19,700.00
Job payback period (in days)	180	150	120
Return on investment (annual)	12.82%	135.64%	45.08%
Working capital situation assessment	EXCEEDED	OKAY	OKAY

No. 2 regarding the percentage of business failures. It is a business famous for what poker players call a *bad beat*. When the signs appear straightforward, you will win; you may be defeated when the "last card" is drawn.

We hope you agree the degree of risk in construction contracts is high. Now we will discuss what a fair charge for that risk should be.

The Reserve Bank of Australia (RBA) is one place to start. The interest rate it charges banks for borrowing money is the "risk-free rate". It is assumed that the funds due in all situations will be paid back; hence, no risk. That rate at present is in the single digits. What is not as important as our business has a risk and the amount charged for the risk of construction work should be above the rate set by the RBA. A double-digit rate is rational.

However, deciding what double-digit rate to charge is not so clear. For example, what value between 10% to 99% margin might be reasonable?

A fair example is to look to credit card companies. They charge a range between 9% to more than 20% per annum for outstanding balances. In addition, many have additional user fees. Finance rates and annual fees represent reimbursement for the risk assumed by credit card companies and input into gross profit calculation. Bankruptcy laws have changed, making it harder to discharge credit card debt, which translates into less risk for card issuers; their reimbursement for risk assumption has increased even more.

Additionally, these companies also enjoy less risk per customer since they have hundreds of thousands, even millions of customers. On the other hand, contractors do not have the luxury of low bad debt exposure since contractors have a low number of clients. The risk that even a single client may default, or even receivables being slow in coming, symbolises a great chance of financial problems for any contractor.

So, we suggest a contractor's risk profile is greater than that of credit card companies. Consequently, we should have a higher return than they do. Hence, ideally, the risk rate charged on a contractor's work should approach 30%, if not more.

However, getting 30% on your investment may be impractical if you want to win work. Or is it? This is a rate that is not a *profit margin,* but a *premium on the contractor's working capital* lent to the project. Again, see Table 5.1 and 5.2.

In construction, PROI is an odd concept. It is one that many people were never taught in our industry's unique economics. There is some complexity to the equation, so let us explain.

First, let us agree that the money tied up to build work is the proper basis for calculation. In other words, we need to charge for *renting out* our working capital. Unfortunately, contractors rarely catch up on project cash flow. Additionally, there are no guarantees on the accuracy of our estimate, the project going well or that payment will be made in a timely fashion. Hence, we must charge a larger percentage than most other industries on the project's financing (lending money to).

There are many other factors, but these are primary in our view. To calculate a PROI on a project, we input the following elements:

- Amount and timing of assets, including cash outflow (uses).
- Amount and timing of cash inflow (sources).
- Gross profit dollars (reward).

Note that the return on investment for a construction firm starts at the project level. Therefore, a contractor's business cannot be robust without profitable projects.

These three factors are used in a mildly complex equation to arrive at a PROI. We unabashedly recommend that construction firms use PROI as a preliminary pricing model. Our research has confirmed that using a return-on-investment calculation at the bidding stage is a "good operating practice". Contractors utilise this, among other factors (such as backlog and competitor's tendering statistical history) to set prices. It keeps them away from bad pricing decisions and overly optimistic projections.

Indeed, you might use this same formula to ascertain a completed project's financial performance return. Additionally, looking at a department's year-end results measured in PROI starts a healthy conversation. These exercises are valuable.

This type of approach has other benefits. As the PROI of a construction firm's projects improves, so does the balance sheet and profit-and-loss statement. There are little debt and cash flow drag on the company. This is an efficient approach.

As an additional measure, return on capital employed can tell you more general financial facts. For example, it will undoubtedly answer the question, "How well my investment in this company is doing?". Critical as construction company shareholders want a high price when they exit. It is measured as a percentage and can be used in many ways, such as 1) proving that new processes are working, 2) assessing the value of a new service offering and 3) supporting a selling price of the firm set by the owner.

Working capital turnover may be also described as the cash-to-cash cycle. ROI in construction is dependent on 1) the cash-to-cash cycle and 2) net profit margin. Know the ways to speed cash flow and better return on investment results.

In conclusion, earning an above-average ROI is a complex endeavour. However, many inputs such as careful client selection, high compliance with good practices, minimal rework and sound financial management may be a single metric.

To be clear, installing great work is the keystone of success in construction. The value of craft skill and project savvy cannot be overstated. Meeting client expectations with a smooth construction process and a trouble-free installation increase the likelihood that value-oriented buyers will use your services again.

To reemphasise, getting compensated timely with less debt and cash flow drag allows contractors to continue to attract good people, partners and projects and to upgrade internal infrastructure. The importance of using PROI for a predictable future cannot be ignored.

Growing Your Construction Business: Scaling Up is "Lumpy"

Construction asset investment is a *lumpy* business. Some inputs to construction comprise significant expenses, and their costs are hard to allocate to particular jobs. Any construction contractor who has tried to improve their process for distributing their overhead costs evenly across jobs has experience with lumpy inputs.

The lumpiness is not true of all inputs. Depending on the market sector, a contractor can buy commodity materials as needed. These inputs act more like a "flow". But big-ticket items like trucks, machinery and buildings cannot be purchased as required. Staff members such as project managers are also lumpy since they generally need to be hired full-time with a long-term commitment to attract the best talent.

When a construction contractor buys a new asset such as machinery or a building or hires a new project manager, this introduces a new risk. In our low-margin business, maintaining disciplined growth is crucial. It is well accepted that contractors who scale up slowly maintain high-quality craft skills, project management, put-in-place craft output and ultimately, client satisfaction. Contracting reputations grow upon timely delivery and many years afterwards as the client utilises the well-built shelter or infrastructure.

A construction contractor who grows too quickly faces much risk and ends up in a suboptimal place construction contracting's risk-reward curve. This section is about making the right choices for fixed asset purchases, including when to purchase a lumpy input and when to pass.

A critical decision point for contractors occurs when the company's jobs outgrow their current lumpy assets. For example, for a plumbing contractor with four trucks in full-time use in four projects, taking on the fifth job may mean buying another yute. For the main contractor looking to expand operations, an office expansion might be required to house additional project managers, estimators and other office staff.

Additional turnover from more jobs behaves as expected: a contractor who grows carefully should see similar productivity levels from the extra project. However, unfortunately for construction contractors, the costs associated with that additional job can be significant compared to its revenues. Figure 5.2 shows a turnover function and a cost function for lumpy inputs for an average capital-intensive construction firm. For simplicity, these lines represent new income and costs associated with additional jobs resulting from the lumpy inputs, not overall turnover and expenses.

In this scenario, the company's growth in lumpy assets is irregular. However, points A to B to C are no problem: new revenues are disproportionate to new lumpy assets. Therefore, presuming that this contractor has sufficient lumpy assets to do the job well at Point C, this contractor wisely left operating space to grow their existing staff and fixed assets.

The problem occurs in growing Point C to Point D. This change might have happened when the contractor bought that new truck or a new building or hired the additional project manager. As we can see, the problem is that even though revenues have increased, costs related to fixed assets have increased even more. Moving from C to D will disrupt this ratio in a way that moving from A to B will not.

Moving from Points C to D can be a risk for a contractor. As any construction contractor who has been around for a few years knows, getting the next job is never a sure thing. Support that extra project manager or justify that new office building with continued work is not guaranteed. In economics, we call this a *random walk*, meaning that the behaviour of markets and agents in a market – and therefore buyers of construction services – is unpredictable. What looks like a logical progression evolves to a series of random steps that have converged on the path. So, naturally, there were deviations and dips along the way. A successful construction contractor has taken more than a few random walks. Those who either grew conservatively or were lucky

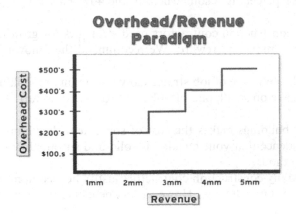

Figure 5.2 New Costs and Revenues Associated with Lumpy Assets of an Average Capital-Intensive Construction Contracting Company.

come out still standing. Those who bought that truck or building at the wrong time ran out of cash struggled to get back in the black.

Growing a construction firm is more straightforward than most other industries. Increased bidding or proposing activity begins the process. Suppose one takes both the value of project bids and the average win rate. In that case, a fairly accurate turnover growth projection can be made, thus allowing for justification for additional staff and assets before the growth occurs.

Even more complex is having the additional high-quality craft skill to grow. Top-quality craftwork is non-negotiable: clients expect this even if the site team is brand new to the company.

A standard metric in construction contracting is *Revenue/Net Fixed Assets*. To be clear, turnover is the contract value billed each year. Net fixed assets are represented by their book value (purchase price less depreciation).

Fixed asset planning, acquisition and utilisation is a subject for another paper; however, some basics should be mentioned here. These come from our experience and research:

1   Before establishing vendors, any purchaser should first seek pricing, terms, conditions, warranty periods and service fees. They have a complete assessment and perspective of the total cost of ownership.
2   The first option is to consider an established product or service that has been in the market for an extended time. In our experience, several leading-edge products over the years were introduced to the market too early and should have continued to be in Beta testing for several months.
3   Their contractor friends should sell contractors. Salespeople do a great job in many ways. Still, those leading and managing the same business know all the issues and can speak creditably about any adverse effects in detail, including essential utilisation planning considerations.
4   Asset newness is a solid indicator of managing the age of your assets, especially equipment. It is the percentage of the useful life of any investment. If relatively new items are a high percentage – it may cost more in payments and interest. Fairly old items may cost more in maintenance and idle time due to breakdowns (unless you are/have a crackerjack mechanic on hand). Most firms are well-advised to keep asset newness percentages somewhere in the 40%–60% range.

Getting back to the main point: construction contractors need strategies for growing sustainably to reduce risk and thus maximise reward. We recommend the following:

*   *Scale-up strategically.* Pick the new type of job strategically when you are looking to take a step from C to D. Take on small, one-off jobs first to test your craft skill and project management.
*   *Rent.* Renting equipment or buildings makes the cost of some items less lumpy. Renting until you have confidence that your backlog is solid and sustainable is a prudent risk management method.
*   *Do not rely on tax benefits.* Many construction contractors justify new equipment investment with expectations of tax reduction. However, tax benefits cannot make up for the additional costs of the assets. Your tax rate is less than 40%, so you pay more than 60% of the cost. A better strategy may be to keep 100% of your funds. The "Rule of 72" will be explained further.

These strategies will yield a more measured growth scenario where the additional costs of new lumpy assets will be balanced by new revenues generated in part from those fixed assets. Figure 5.2 shows a more measured investment pattern that leaves time for new turnover to catch up with the new costs of lumpy assets.

Asset purchase planning reflective of Figure 5.2 is critical for all contractors, especially in a capital-intensive business such as Heavy or Highway Construction. However, sometimes an asset can be paid for from one project's use. It is exciting to think that after a single project, the cost of that asset's production is much reduced (maintenance and insurance, with operating expenses).

In other construction contracting sectors that need little in the way of fixed assets, this step cost dynamic is not a substantial consideration. However, there are "de facto" fixed asset decisions, such as hiring a full-time and long-term project manager or estimator. These talented people in our industry's well-documented shortage have many options and other offers. However, their cost and their value are high; thus, onboarding and duties must be planned well.

Of course, management staff can be hired as a contractor wins projects. In adding an estimator, the desire to increase projects tendered should lead to more projects won. So, there is no "chicken and egg" problem. In most cases, there is an increased bidding activity which foreshadows growth.

As an aside, from a financial perspective, it is prudent to have at least six months of any asset's owning and operating cost saved before purchasing. But, again, this is a disciplined method of attempting to keep debt out of a construction firm.

In summary, fixed asset acquisition is challenging when growing a construction business. It can be significant in a capital intensive one. However, all construction firms must deal with "de facto" fixed asset acquisition, such as talented managers and supervisors. This means that all construction firms must plan acquisition and utilisation with care to experience increased productivity.

In one way, lumpy asset dynamics show the demands of growth. What is important to remember is that growth is a choice in construction contracting. We know plenty of owners of small construction firms who are quite happy and those who are majority stockholders of large contracting organisations. Smaller contractors seem to enjoy a simpler lifestyle. Those who own a considerable concern are rewarded each day with challenges and accomplishments. Each type of contractor has made a choice. They know growth is not a requirement for any contracting firm to be profitable.

Managing business growth is a challenge. Being aware of the issues before executing such a plan is critical. In construction contracting, superior planning and execution are rewarded with lower cost, higher productivity and profits, whether in a project, business, or strategic setting.

The Rule of 72

Understanding financial management is critical to the success of any construction contractor. Harold Geneen, the late chairperson of ITT, stated it succinctly by creating the ten business rules. To paraphrase:

"The first rule is don't run out of cash, the second is don't run out of cash, the third is don't run out of cash and rules four through ten are not that important".

Financial management's importance cannot be overstated. The Rule of 72 is a crucial concept to financial independence. Maybe it should be the 11th rule.

Which is more? Five million dollars or a penny doubled every day for a month? Surprisingly, the answer is the penny (maybe not a surprise if you were a good math student). This illustrates the power of the Rule of 72. This is a rough but straightforward financial calculation that dramatically shows the power of doubling. Moreover, it demonstrates the power of time with interest to produce a large financial outcome.

As a practical example, contractor employee A invests $2,000 a year for six years and then does no more investing for 39 years. All the while, contractor employee B does the opposite. they do not invest for those first six years but invest $2,000 a year for the last 39 years. So who will have more money in the end? Contractor employee A.

In the past century, many have said it and we believe it – the most potent financial action in the world is the compounding of interest.

The math of this rule is simple. It gives us a calculation to determine the doubling period for a given interest rate. For example, if a principal earning has a fixed rate of interest of 6%, that sum will double in value in approximately 12 years (72 divided by 6 equals 12). On the other hand, if the principal sum is earning interest at a rate of 12%, it will take only six years for the principal sum to double in value (72 divided by 12 equals 6). Divide other interest rates into 72 and observe the time required to double your initial principal.

Indeed, this is based on having money to invest. Collected profit is the largest source of funds in the construction industry. This solidly points to the importance of expense discipline and financial management for construction firms and their owners. Done with frugality and care, the reward is substantial wealth at a relatively early age-otherwise, we work longer than we want to.

Using the same logic for expenses provides us with an insight into business costs. Let us review the example of a typical tax write-off; of a work vehicle used in a company's business. The Australian Government may institute a significant tax incentive for purchasing a new vehicle in a poor economic climate.

There are certain tax deductions in Australia for business depending on the year due to a recession or not. However, the immediate write-off of gross taxable income in capital investment is expected in poor economic times.

As a result, some contractors (typically younger) have purchased expensive yutes and other capital equipment. They assume saving taxes is one of the best financial strategies. Is this the wisest choice? From an economic perspective, could there be a better alternative? Hint: many societies have disliked the tax authority for over 2,000 years.

Let's use the example of making $40,000 net profit before tax. We will assume that we will not make this purchase, keep our old truck and pay the taxes. What will be the outcome?

For the sake of our discussion, our example firm is a sole trader. From our earnings last year, let's assume we will pay the government around $10,000 of taxes, leaving us $30,000 to invest.

Our $30,000 is for our personal use. We decide to invest it in a reputable, no-load share fund. Without hesitation, we can project a 10% return on this investment. Using the Rule of 72, we know in 22 years, this sum should double three times and grow to $240,000. It is exciting to think of the possibilities, such as repeating this net profit scenario and investing it likewise over ten years.

Table 5.4 Comparison of Two Tax and Investment Scenarios

Year	Keep the Old Truck, Pay Tax and Invest	Buy the New Truck, Save the Tax with No Investment	Comments
20XX	$30,000 in investment but pay $10,000 tax [*]	$0 in investment and pay $0 in tax, but new truck using accelerated depreciation [*]	Wealth gain while paying tax versus purchase of a new work vehicle while avoiding tax.
20XX + 22 years	$240,000 in investment portfolio from year 20XX [**]	$0 in investment portfolio from year 20XX	Consecutive years of this scenario represent an opportunity to increase investment wealth.

Notes:
[*] This is a hypothetical tax scenario. Please consult your tax advisor.
[**] Projected 10% gain each year for 22 years is hypothetical and may not occur.

Take the opposite example and you buy the new truck. What is it worth after 22 years? Did it produce $240,000 more in net profit than the old truck during its useful life? Was its savings in productivity that much greater? (Table 5.4).

As an important aside, with the latest flood of increased government entitlements, stimulus spending and other unfunded giveaways, we will see inflation in coming years. One way to combat this economic problem is to be an equity or stock investor. We see this for working professionals as well as retirees. Strange as it may sound, retirees may be better protected by continuing their diversified stock holdings past their working years. The ethic of investing in bonds in retirement may not be the best strategy. Inflation will decrease the value of those earnings, which are smaller than stocks, to begin with. Company earnings keep pace with inflation. Stock prices and their dividends are mostly a gauge of those corporate earnings.

The working number for retirement income is 4% of the principal. Whatever you have saved, can you live on 4% of that number? Consider all your income streams, but only budget 4% of your nest egg or investment value. This may shock you, but remember, we expect to live much longer than our parents; we certainly do not want to go back to work when we are 85.

In our industry, maintaining trucks, equipment, tools and other assets well should take on new meaning in light of this rule. Even if society does, we cannot have a careless, "throwaway" attitude. It would undermine the very financial security most businesspeople seek for themselves and their families.

To restate, if we are less concerned with paying taxes to the government and more concerned with our wealth, we can build our long-term financial strength. Thus, we are justified to deny ourselves the extravagance of a luxury SUV or new work truck. In another way, the Rule of 72 shows few worthwhile tax incentives. What we want are tax credits. Sales professionals may imply some sort of financial advantage through a tax write-off against gross income, but we disagree. At any rate, any significant business investment needs to be reviewed closely for utilisation and productivity contribution.

Take this economic rule of thumb seriously. Consider it during your next tax cycle and in your business plan. At the end of a career, we believe that being financially independent is what most contractors and their staff desire. Understanding the Rule of 72 is a primary step to getting there.

A New Look at Job Cost Reporting in Construction Contracting

A sound grasp of financial management is crucial for any contractor's must-have education. Since the inability to pay bills is the fundamental reason for bankruptcy, it is at or near the top of critical topics.

"My father was the first fiscal conservative I ever knew. He was the first of many things for me. But he also had a devastating sense of humour. I appreciate that now, as I appreciate the value of a light moment. Humour is critical for our mental health. Early one morning, he said, "There is one thing money can't buy: 'poverty'".

What was meant was that money gives you many business options. Therefore, we need to have a laser-sharp focus on finances if we have financial options now and in our later years.

The Cost of Construction Contracting

What causes construction turnover to change or costs to behave? The answer is not to be found in financial statements. Instead, it is located on-site, in the client's office and the contractor's operating methods. However, unless you use the mattress or coffee-can method, your bank account tells the story of your executive leadership, including your financial management skill.

Before you can get an accurate reading of what is happening in your business, you must collect and analyse accurate data that comprise the picture of your financial health before you can make sound financial decisions.

In the construction contracting business, finance and accounting perform a reporting function. As financial and accounting managers' report to senior management, it is the role of senior managers to act on the information provided. This is financial management. Construction being the unique business it is, you must make several detailed considerations before making any decisions.

Simply interpreting positive numbers well and negative numbers harshly is not sound business acumen. The business is construction – not finance. Just as we know that the business is half people, not all people. Construction is a vastly different environment from any other in which most professionals have ever operated.

Your job cost process must be considered a sacred part of your firm. If the projects are profitable, so is your firm. But, of course, the reverse is also true. Sins such as hiding costs in unrelated cost codes and overwhelming the site with hundreds instead of dozens of cost codes should not be tolerated. Misinformation here is not what you need.

Your job cost report should not be an afterthought. It can be a leading indicator of risk or adverse.

events. It must be a predictor of "best-case" and "worst-case" scenarios. As your site will not tell you everything and you cannot camp out on the job site, this should be an objective assessment of what is occurring on your projects.

Jobs can generate profit and reputations. All roads go through job turnover and their expenses from a financial perspective. Contractors must be experts at the financial management of jobs because if jobs are profitable, then a construction firm can be too. There is no other way. Understanding job costing in depth is mandatory for anyone who sincerely wants to create a substantial and lasting contracting organisation.

Job cost reports are an essential piece of management information for managing construction projects. The results of this report flow into the Profit and Loss (P&L) and Balance Sheet Statements. Superior job cost results make a construction company's financial health strong.

From our perspective, many contractors misunderstand their purpose, structure and content. Therefore, we will share below our explanation of one way to organise them and interpret the results.

Regardless of the organisational method you choose, however, to report well, we must account well. Therefore, the accurate capture of numbers is critical. While job cost reporting is crucial in the accounting process, we view it as the weakest link because it accurately captures the most difficult data.

We know that the people on-site usually do not have accounting degrees or bookkeeping certificates. They are not hired for their financial skills. We all tend to find this to be the least fun part of the job. Many people lose accuracy when collating, sorting and submitting cost, unit, time and other numbers. Nevertheless, detail and accuracy are crucial in all phases of data capture for accounting and finance functions.

So, a standardised numbering system and descriptions in the Job Cost Report are easier to learn. Limit the number of cost codes and use long descriptors to limit confusion. Ensuring simplicity and clarity of meaning results in increased accuracy and efficiency.

Therefore, to be practical and effective, we should keep our job cost capture system straightforward. This ensures that the people responsible for recording the numbers can do this aspect of their job well and will not be adversely affected by confusion or tedium.

Job Cost needs to be captured accurately for another reason. You will need to have a clear picture of the cost per unit at some point. There are critical people in the construction arena who need this information:

- **Contractor:** will need to have their facts straight, not just opinion, when coaching or correcting poor supervisory behaviour.
- **Chief Estimator:** will need to know what "break-even cost" is in competitive bids or negotiations.
- **Client:** will need facts when challenging your variation or claim.
- **Bank:** an essential partner to the construction in qualifying for many types of work. In our opinion, project owners that place a high value on a contractor's financial standing so it facilitates less competition. In addition, banks want contractors who know their costs well.
- **Site Supervisors:** This is a large part of their scorecard. It tells them if they are doing a great, mediocre, or poor job.

Your job cost report should not be an afterthought. On the contrary, it is a leading indicator of risk. As your site staff will not tell you everything and you cannot camp out on the job site, this is a critical objective assessment of what is occurring on your projects that warrants severe planning and implementation.

The first rule of Job Cost Reporting is that it should be a *predictor* of loss or gain. There is no other reason to generate them. There is plenty to manage without additional paperwork. As many have said, "Manage by exception". Looking for figures that deviate from expectations is a good practice. Be vigilant for any signal of potential loss or financial reverses. These are reasons to inspect and then to act. It takes approximately four to five profitable jobs to break even for one bad project. Therefore, having a proactive job cost process is a crucial success factor.

Job Cost should be a sacred part of your firm. If the projects are profitable, so is your firm. The reverse is also true. Do not tolerate sins such as hiding costs in unrelated cost codes and overwhelming the site with hundreds of cost codes. Again, this is a critical piece of information for you to manage effectively. Misinformation here is not what you need.

Job cost reports should help managers quickly understand what is happening on projects. Remember, the faster you complete the business cycle (*work acquisition – install work – get paid*), the better you will make your cash flow, the more you will increase profit and the more stable and certain you will make your company's future.

Many ways to represent job cost data exist. First, we will review an example of a job cost format and protocol that we recommend starting with (see Table 5.5). Then, you will evolve this into a better system for your firm.

We will analyse this report, explain why this format and why we believe its calculations are desirable. Then, we will review starting from the far-left column and proceed to the right-hand edge.

Column 1. Task #: Assign a unique identifier for each task; limit the number of tasks assigned to each supervisor. As a rule, fewer is better. Since most site supervisors do not have an accounting degree and do not desire an administrative position, be careful about assigning too many detailed cost account numbers for them to track. Typically, the more we delegate, the less accurate is the information obtained. We recommend fewer than forty unrelated tasks for any site supervisor who is supervising direct labour. "Pencil whipping" or quickly writing an estimated amount without thought is likely to happen. Assigning fewer numbers with plain descriptions results in more accurate and reliable cost accounting. Forepersons are in the best position to witness work. Give them a practical, easy to understand the way to record hours and other costs accurately.

The company should standardise these numbers and should not change them from project to project or year to year. This practice reduces the likelihood of errors and allows for increased speed and efficiency as your experienced staff becomes more familiar with the reporting system.

Column 2. Description: Label each unique cost item in an easy-to-understand and remember description. Be sure to use ordinary language; avoid jargon that is too technical. This will encourage more accurate cost recording and fewer calls to the office asking for clarification. If your software package allows for a 55-character description, use it. The clearer the label is, the less confusion. As with cost codes (Column 1: Task #), these cost item descriptions should be standardised by the company and should not change from project to project or year to year.

Each work item should be an assembled piece, a completed item that is easily measured. So, the task # and the description represent constructed walls, air duct, curb, pipe and the like. This is so that the project manager can measure physical completion each month without great effort. Typically, a steel tape and a set of plans allow measurements to be completed and then input into the job cost software.

Table 5.5 Job Cost Report Example

Job Cost Report

Task #	Description	Estimated Unit Count	Actual Unit Count	Per Cent Complete	Estimated Unit Cost	Actual Unit Cost	Total Cost to Date	Best Case	Budget	Worst Case	Task Complete
6–100	Underground - 100 mm	1000	995	100	$3.55	$3.99	$3,970	$2	$3,550	$440	Y
6–105	Underground - 200 mm	1250	1260	101	$5.44	$5.60	$7,056	–$2	$6,800	$200	Y
6–110	Underground - 300 mm	1600	1601	100	$6.98	$7.01	$11,223	$0	$11,168	$48	Y
6–115	Overhead - 100 mm	2000	2010	101	$4.65	$4.85	$9,749	–$2	$9,300	$400	Y
6–120	Overhead - 200 mm	1450	750	52	$6.90	$6.80	$5,100	–$145	$10,005	$0	N
6–125	Overhead - 300 mm	1700	850	50	$7.66	$9.88	$8,398	$1,887	$13,022	$3,774	N
6–130	Final Connect - 100 mm	1100	222	20	$5.20	$4.75	$1,055	–$495	$5,720	$0	N
6–135	Final Connect - 200 mm	1260	145	12	$6.48	$6.80	$986	$357	$8,165	$403	N
6–140	Final Connect - 300 mm	1640	355	22	$8.03	$9.00	$3,195	$1,246	$13,169	$1,591	N

Column 3. Estimated Unit Count: This entry is copied directly from the estimate. It is a number you input, not a calculation. This is a core element of this job cost protocol. Beat the budget, and you make a more significant job profit. It is that simple. It is a critical element that delivers a solid understanding of a contractor's schedule, cost and physical completion. The estimated and actual count gives us critical information during the project and after it is completed.

Column 4. Actual Unit Count: This is the exact count of work completed by the date you complete the report. For example, after week one, you may report that ten units of the estimated underground piping have been installed and after week two, perhaps 60 units will be the number entered here. This number will change as work progresses. Using easily measured units allows the site to count and project cost to complete often (weekly, daily).

Comparing this entry to the Estimated Unit Count is critical in three ways

1   Construction allows a completion percentage to be roughly calculated, thus helping derive a final cost at completion.
2   It helps determine any over-or-under-billing for financial reporting purposes. In addition, this is a clear, efficient way of calculating – meaningful in your bank's questions, surety, or other partners.
3   This method is logical to understand if a cost overrun is caused by the estimating problem or site mistake. Undercounts have a direct effect on the cost estimate. But, of course, overcounts are welcome, making cost estimates higher and likely converting into added profit dollars.

Column 5. Estimated Unit Cost: The estimated unit cost is also input directly from your estimate. It would be part of the budgeting as you set up the job to be built.

Compare this number to the cost records of your company. This estimate should align with actual costs on other profitable jobs and current market conditions. If your assessment is reasonable (and the counts are correct), but the actual project expenses are higher than estimated, this indicates a site problem. On the other hand, if the unit cost is the lowest ever recorded, it means an estimating error (over count) unless you can discover actual circumstances that caused a significant cost variance in the project.

Estimated Unit Cost numbers should be medians (where 50% of previous cost datum fall above it and 50% fall below it), not averages, of the cost history of the company. We all have good days and bad days. The productivity of these days varies widely. The median dismisses the extreme occurrences and derives the average and expected unit cost number for most days.

Column 6. Actual Unit Cost: This number is calculated by dividing total labour cost by total units completed. Compared with the Estimated Unit Cost (column 5), this number contrasts the Estimator's assessment of the work with the Site's performance.

Important Note: Columns 3–6 comprise the critical data needed to understand whether a project overrun is a site or estimating problem. A report displaying a reasonable Estimated Unit Cost with a lower actual cost is to the Site's credit. Not surprisingly, if the site has higher than estimated costs, that indicates an issue. Overall, these columns and this report act as a tripwire that sends up a flare signalling trouble. Investigate the details only when extraordinary things are happening – good or bad.

Column 7. Percent Completed to Date: This is the percentage of physical completion of the project. Completing the project with excellence is the primary goal. The amount of work installed allows you to know several things, such as the amount earned versus the amount billed to the client. Therefore, you can quickly see if you are overbilled or under billed. Also, at year-end, you can more efficiently compile your Work in Process statement.

Column 8. Total Cost to Date: This calculation is from site records. In this report, the entire site cost is first inputted from cost records. From there, the Actual Unit Cost is a function of Total Cost divided by Actual Units. Again, the total cost to date is an accumulation of all invoices, bills and other Account Receivable on the job and then we divide through by the units completed. This backwards calculation is the only way to arrive at a unit cost quickly.

Column 9. Best Case: The Best Case is the optimum result as projected by the job to date performance. In favourable situations, the calculation is the total current performance to date plus the remaining work performed at the present (superior) rate.

For projects showing an overrun, the Best Case is the total current performance plus the remaining units times the estimated unit costs. It is an optimistic estimate.

Given that the project to date performance is flawed, it is not responsible for suggesting that the task will finish within budget. Alternatively, in cases of better-than-budget performance to date, the excellent unit cost performance is multiplied by all the units. This will undoubtedly show savings in the budget.

In both best-case and worst-case, we recommend using dollars (or your native currency). We discourage percentages since this is not tangible. On the contrary, showing a loss prompts company staff to action.

Column 10. Budget: This is a statement in dollars of the task estimate cost. It is a straightforward calculation derived by multiplying the estimated unit count (Column 3) by the estimated unit cost (Column 5). Again, your budget must be based on assembled work units that are easily measured to update the actual expenses each month compared to the estimates.

Its place at the end of the job cost report allows the contractor to quickly compare the budget to best/worst without using a ruler or straight edge.

We suggest that Job Cost Reports be completed each week. The shorter the time monitored, the more valuable it. Risk increases when issues are unknown. Problematic line items do not have time to fester and test your firefighting skills.

Column 11. Worst Case: The Worst-Case is a derivation using the least optimum numbers. This can be thought of as the most conservative calculation.

In overrun situations, the calculation is the total current performance to date plus the remaining work performed at the same poor performance rate (it does not matter what remains to be installed). The result is a higher number than the budget.

When a task is beating the budget, this will multiply any remaining units by the estimated unit cost. The result is a worst-case less than the budget.

Column 12. Task Complete: This is simply the manager's report of whether the job line item is complete. No further cost changes may be inputted once a "yes" is entered into the cost report.

As people build and supervise, they exceed cost budgets. Subsequently, there is a temptation to hide costs when a cost code is running over. Charging labour expenses into unrelated cost codes is a common occurrence in construction. Verifying when a cost code is complete will minimise the risk of caving into this temptation.

We must be cautious when using these methods. There are some traps to using any approach. Nothing is ever a perfect solution. All approaches have unintended consequences. Nevertheless, we can offer some insight based on our experience that may help you with this process:

- Make the cost codes small in scope. List more prominent aspects of a project into smaller segments that are easy for all to understand and for supervisors to observe. For example, instead of listing "site clean-up", list more discrete components such as "third Floor including Stair Clean-up" or "Culvert A to Culvert B Dressing".

Breaking the tasks into discrete, manageable units will also help you keep costs codes in the right categories. This will significantly assist in budgeting and accounting functions.

- As you know, approximately the first 5% of the time on any job involves moving materials, getting crews lined up and organising the work. Therefore, at the start of any project, labour productivity will be lower. As a result, your report will reflect a higher unit cost than you estimated. (As a suggestion, you may want to have a mobilisation cost code for each. This takes away the excuse of a slow start).

At some point in the first half of the job, your crews will gain their stride. So, we must ask, at what point (or percentage of job completed) we should start to be concerned if the unit costs remain higher than the estimate.

In our experience, you should begin to see your labour costs matching your estimates when approximately 10% of the task items are installed. Your site supervisors will be better able to judge these factors as they gain experience. However, any discrepancy beyond this point is a reason for further investigation.

- Giving your site supervisors a quick method to check daily productivity and compare it with the estimate is effective. It forces the site supervisors to think more like managers. As they see performance measures in their daily metric, they will act accordingly and keep costs managed.
- Most project managers and site supervisors find it easier to think about time and work units than money or percentages. Therefore, in your site reports to your site team, hours should be one of the measures. For example, if you are asking a site supervisor what percent of a complete work task is not helpful. Pride and other emotions may colour the answer.

A better question is: When will that work item be complete or how many will be completed by the end of the week? Most site people think in terms of time or physical progress. The question is clearer and therefore, the answer will be more accurate.

Job Cost Reporting is a "good operating practice" in the construction industry. There is no debate that it is a critical one for construction contractors. However, how it is performed is another question. Sometimes it is not performed optimally. We suggest that a predictive element is crucial. This may force some estimators to change their process and capture assembled units. Such is the price of improvement.

Alternatively, we must recognise that some solid financial practices in other industries may not be helpful in construction. That discussion is for a later date. Sometimes the newness of an idea captures excitement but is its only value. Sadly, a contractor might find out too late in time and dollars.

Be sure that any financial practice you consider using has proven efficacy in construction. However, you must be careful since money in the bank is the only barrier between you and bankruptcy.

Compensation Plans: Cultivating an Owner's Mindset in Employees

Most owners would like to see their employees care as much about the company as they do. In our low-margin business, every mistake counts against and saving contributes to the bottom line. When employees understand this, everyone wins. Thoughtful compensation plans also incentivise people to work harder than their peers and be paid for that hard work. At the same time, compensation plans represent an opportunity for employees to share in the year-to-year improvement of a company. We think the goal should be to craft a compensation plan that incentivises successful behaviours and top quartile outcomes.

The size of a paycheck can be a highly emotional subject for many people. Some see it as a personal scorecard and others as a means to an end, i.e., purchasing a house, truck and education. This is obvious but critical to understanding. To apply this issue of heightened emotion in compensation is to understand two things 1) Companies must be careful in all salary and bonus calculations and discussions 2) These same firms must meticulously test any plan against all possible situations. To say it differently, changing the plan in its first year or very often throughout its life will make the company committed to paying for performance seem less serious. The employees may adopt that attitude.

Construction contractors can craft compensation plans to incentivise employees to think like owners. However, other times, they may unwittingly incentivise employees to be lax with details. To paraphrase a message one of our clients aptly posts on their office wall, "Every system is perfectly designed for the results it delivers". The purpose of this section is to help executive teams create *construction compensation plans* which structure the right mix of incentives to help employees think like owners, i.e., each company dollar saved or earned translates into a greater personal financial reward.

Overall, when employees make contributions that add to the company's average bottom line, they should be rewarded. This means that each dollar saved over a minimum should be part of the bonus plan fund. Average or previous year performance should not be rewarded significantly.

In construction contracting, a properly devised incentive plan may be complex in its creation but simple in its execution. All compensation plans should have three parts: 1) Base salary or wages 2) Bonus or incentive compensation 3) Benefits. The market typically sets the amounts, i.e., a company's competitors, as recorded by industry surveys. Furthermore, each of the three components should be independent of each other. That is, each part should **not** grow or shrink in lockstep due to a change in any element.

Furthermore, how each component is computed should be transparent and regularly shared with employees

With the long view in mind, a good compensation plan may enable employees to think about and subsequently purchase company stock when the current shareholders transition to retirement. In this and other ways, a compensation plan should follow a company's strategic plan.

Base Salary – The Start of the Bonus Calculation

To put it simply, the economic value of an employee is what you pay them. So, it should follow that a lower paid person with comparable performance to a higher paid person should not receive a more significant bonus. The reverse is true.

The best compensation package starts with a base salary that is slightly lower than the industry average (and bonus opportunity higher). From industry salary data, we can determine the market value of labour for salaried and hourly positions in various sectors, e.g., heavy civil or mechanical and in multiple regions, e.g., Southwest or Northeast. Bonus determination is partly contingent on an employee's base salary and responsibilities, so it is vital to get the base salary right. We believe firmly in specific construction firm salary market data. It helps set an appropriate level and a total package that is on par with your competitors. Significant especially in new areas such as a Building Information Modelling Director or ones that are high ranking such as a Chairperson of the Board.

In our research, the core reason for a compensation plan is to give employees an incentive to think more like an owner. Each dollar saved or made over a minimum should fund the bonus plan. Average performance must not be rewarded in a significant way. The average industry increase may be a place to start. Improvement and profits incentives make people and companies better.

What is also interesting is that in most bonus programs, employees do not share in losses i.e., a negative bonus, although no bonus is certainly attention-getting if a bonus was paid the previous year(s).

Each of the three components should be independent of each other. Each part should not grow or shrink due to a change in one or another element. National industry data is the best place to determine where to start in deciding compensation packages for specific salaried positions.

We have observed that when a person is grossly underpaid for their work, once they find out, the relationship with their supervisor is damaged, sometimes beyond repair. Do not take that chance. Address those disparities. The result is a fairer and longer-term relationship with all your employees.

Bonus determination is partly contingent on the base salary of an employee along with their authority and responsibility. In many plans, senior executives should assign a factor, say one to four, denoting employee authority and accountability. A designation of one may be a person at entry-level and four is an executive. This assists in the mathematical calculation of the bonus amount. As an aside, authority and responsibility can also be thought of informally as "potential stress level'. This is a good guide for assessing someone's value, i.e., taking the stress off the company owner.

Bonus Funding – Savings Create the Pool

All construction firms need to create a bonus pool not from average performance, meeting the estimate or yearly corporate budget.

The funding source is how we allocate dollars to be distributed. A great bonus plan starts from zero dollars each year and the bonus fund should be added to as a company crosses the line into above-average profitability. The more profitable the company, the more significant the percentage is contributed to the bonus pool. One common way to populate the bonus fund is for an increasing proportion of profit allocated as company ROI moves past 30%.

Many times, the company will find its funding source in the same outcomes as its distribution mechanism. This is normal and is a sign of *alignment,* a powerful unifying effect for employee effort, enthusiasm and collaboration.

Bonus Distribution – Company Values Are Expressed in Payments

The distribution mechanism is how we reward underperformers and overachievers, i.e., allocate the money from the bonus pool to each employee. The distribution mechanism is based on incentives and should reward good performance. The following components might be considered in the distribution of the bonus fund:

Safety – All incentive programs should start with perfect safety. This is the correct focus in many ways. Compliance to safety process adherence is easier to ascertain since studies and systems are more available than ever. A straightforward procedure for rewarding safety compliance should be part of a transparent distribution plan.

Productivity – Human resource utilisation focus is essential for most construction professionals. Compare this to an incentive to capture all variation revenue. If profit management is a significant motivation, then an aggressive capture of all changes, including marginal ones to the project, will not be a problem. However, in our experience, including aggressive capture of scope changes might not improve employee performance overall and long-term. Human resource (and equipment) productivity is a better long-term focus.

Project profitability – Construction project performance should be part of the bonus mix. If people do well in an overall bad year, some financial recognition is rational. Remember, your competitors do not have the same bad years as you do and may pay a better bonus in some years than you.

Adherance to processes – Like safety, adherence to standardised company practices (on-time, complete and done once) is not as hard to measure these days. Additionally, we know that when employees comply with efficient processes, this means 1) less loss due to rework, 2) equipment and materials are on-site and in position when required and 3) reports are done on time. All of this adds up to more profits. Therefore, good processes should be established and shared with all employees and compliance with those processes should be measured transparently. This may take a little work, but the rewards can be profound.

Incentive Structures

Site and office employees have different incentive structures, but the company's overall incentives require the office and the Site to be aligned in their goals to company ones. For example, see Tables 5.6 and 5.7.

One primary goal of the compensation program should be to incentivise the office and the site to work together, rather than blame each other for

Table 5.6 Funding and Distribution Framework for a Bonus Plan

	Funding of Bonus	*Distribution of Bonus*
Construction Project(s)	Results of Project 1 Site 2 Office	Behaviours and Results of Employees
Company Performance	Results of Company	

Table 5.7 Typical Bonus Plan Focus by Type of Contractor and Employee Team

		Employee Location	
		Site	*Office*
Type of Contractor	**Specialty**	Focus on Safety, Schedule and Productivity of Labour and Equipment	Focus on Safety, Schedule and Return on Investment
	General	Focus on Safety, Schedule and Quality	Focus on Safety, Procurement, Schedule and Quality

lower-than-target performance. Site-Office conflict is a significant contributor to inefficiency. Other times, the financial manager is not supporting the project managers and site supervisors with their best effort. Which position is more critical to completing projects on time, with safe and efficient results that should be evident to any executive and only captures and reports results? The incentive plan should follow this logic.

Construction projects should be part of the bonus mix. Some financial recognition is rational if people do well on their project(s) in an overall bad year. Remember, your competitors do not have the same bad years and may pay a better bonus at times than you. This may convert into a short-term recruiting advantage for them.

Specifically, we have found great results with subcontractor site supervisor's bonuses centred on only two of five direct costs, 1) labour productivity 2) Equipment productivity. This focus generates savings in a project's two greatest variables (significant chance of cost overruns).

Overall, cost focus is critical for profitability. Our industry's net profit is 4%, so a steady, error-free and high compliance culture appears needed. If so, an incentive plan based partially on that, then it should have positive effects on projects. However, if a program is directed more on profit, then variations, additives and deductives may reward their capture. We all know variations are arbitrary (luck). Contractors make a profit on either. They may add to bottom-line profit but do not require operational excellence in and of themselves.

If total profit is an incentive to capture all variation revenue. If profit (and not cost) management is a large motivation, then an aggressive capture of all changes, including marginal ones, will follow. In our experience, aggressive capture of scope changes is problematic to growing a relationship with a client. It must be balanced with customer satisfaction.

Transparency, Consistency and Construction Appropriate Feedback

Many construction contractors find that they may not quickly get employees excited about an incentive program in the first year. Therefore, we encourage owners struggling to get employees interested to develop a feedback system that is frequent and transparent. This takes advantage of the "learning and forgetting" cycle that humans are prone to.

People generally remember and comply with processes to a high degree immediately after a reminder and our compliance declines over time until the next reminder. Having quarterly or monthly updates resets this cycle more frequently, increasing good behaviour. However, once checks are paid to well-performing employees at the end of the first year, this concrete reminder is well remembered.

Also, any evaluation system (with which an incentive program is based) should be written in our industry vocabulary and use construction-centric concepts. Unfortunately, it appears that general business language has crept into some yearly evaluation forms of employees. As an example, "Planning ability" is too broad. This leaves some interpretation to the evaluator. A better evaluation language might be "Employee plans their project work four weeks ahead each week".

The Discretionary Component

In addition to the distribution mechanism, an effective bonus plan also has a discretionary component. This reflects the judgment of an employee's performance by their supervisor. This enables an owner to reward behaviours and performance that sit outside the bonus system scope.

The discretionary part of a bonus plan is necessary. It instils fairness. Sometimes people are lucky or unlucky and that must be addressed. If not, luck might be whispered amongst employees as the truth about bonus determination. A small incentive payment encourages a snake bit employee to keep working hard and smart in a poor year. This discretionary bonus keeps the culture of "controlling the controllable".

What-If Planning

A company's condition changes each year. We have never witnessed a company staying in precisely the same place for two years in a row. There must be a test of the system before the announcement to all the employees and the launch of the new program. Changing the plan in the first year due to a lack of what-if planning is a grievous error but controllable. So, taking the time upfront will save embarrassment (and preserve creditability) later.

Typical questions might include:

- What if the company stays the same size or profit level? (Will there be the same desired enthusiasm?)
- What if the company grows enormously? (There will be many new employees rewarded by the system, say after six months?)
- What if the market price for materials rises or lower considerably? (Does it result in a windfall or hardship in the bonus calculation?)

- What if the company downsizes dramatically? (Employees who are leaving involuntarily but earned some bonus. How do you recognise them?)

Other Issues

Smart owners know that there are many considerations in crafting a bonus plan. Some defy categorisation. Here are some various factors to consider:

Employee desires change as they become older – An older employee may have more incentive to work efficiently, such as increased financial responsibilities or a desire to improve their retirement savings. Employees may consider buying into a company stock purchase plan later, whereas they may bypass the opportunity when younger. Always make sure to keep opportunities alive for employees as they move through the phases of their careers.

The long view should be rewarded – Executive and senior management bonus payments may be staged to pay out over three or more years. This keeps incentives aligned with the long-term viability and productivity of the company.

"Mental Owners" will stay with the firm longer – Vital employees may wish to branch out on their own. A well-crafted incentive plan that rewards employees with an owners' mindset can help a company retain its talent. An employee who feels a strong sense of mental ownership and loyalty will work harder for the company.

Poor bonus year – Payments to employees will not grow each year consistently. In any year, bonuses may be minimal. This eventuality must be planned for; otherwise, employees may "push" savings and billings into next year and "pull" direct costs and overhead expenses into this year. Again, this problem emphasises the value of a discretionary component.

Bonus gamers – Some employees may start "gaming the system", i.e., do things that raise their bonus dollars but that do not follow the spirit of the incentive program. This is an excellent reason to have a discretionary part of the bonus program. Additionally, an executive should consider planning to modify the program every three years.

In summary, bonus plans can make working the long, hard hours required in construction worthwhile. In addition, if structured and executed correctly, they can help people work with an owner-operator mindset. Of course, no bonus plan is perfect and some criticism by employees will occur. However, a good plan takes the mystery out of what behaviours and results get rewarded.

When in doubt, pay the bonus as calculated by your system. As each of us knows, almost all employees want to do a great job, do not give them a reason to lose motivation and develop a poor attitude. If there is a question, those employees who possess marginal habits and thinking "will come out in the wash" over the next year (s)-the positive ones will also.

Once a bonus system is part of company culture, it is not a significant change for employees to purchase company stock as the owner(s) start transitioning towards retirement. Internal buyouts funded by "earn-outs" are common in ownership transitions. Those valuable and earnest employees who want to own your construction firm can do so due to the mechanism of a bonus program.

If you want the owner mentality to pervade your employee ranks, a thoughtful compensation plan including a bonus program encourages that mindset. However, it must be devised with critical mechanisms and considerations. If done well, the construction firm owner(s) will not be the first to arrive and the last to leave the office or the job site.

35 Ways to Reduce the Expenses of a Construction Firm

Across Australia, contractors will "hit the wall" of minimal cash in the bank and project prospects. Some companies experience this in the early years. Time is needed to assess the situation. However, if you are a young business, would not a list of actions be helpful and proactive not to have financial stresses?

The first rule of business, "don't run out of cash", takes on a new urgency. As you fight this battle, we offer 35 ways to cut expenses. Since there is no silver bullet in construction and it is a 4% net profit industry, doing the little things will add up to the big wins – staying profitable, cash flow positive, and most importantly, actively in business.

1 "What gets measured gets managed". As a start, measure waste, planning, forecasting of critical resources and compliance to processes. Place particular focus on labour since it is our highest overall cost and has a ripple effect on all expenses.
2 Pay employees mileage for the use of their vehicles, while keeping "Gray" travel minimised and disciplined.
3 Use trailers instead of vans or pickups. At the extreme, livestock or horse trailers have a utility.
4 Send an income declaration to bad debtors. This may not force collection, but the ATO will feel better and know about the value received. In addition, it may prompt a phone call and a beneficial conversation from the offending party.
5 Be ruthless when it comes to collections, even if that means sitting in someone's office to collect money. Old retention collected can mean a new life. For example, ordering a pizza for lunch in earshot of the client sometimes prompts payment.
6 Ask your service providers such as cell phone, permit acquisition, or trailer rental to analyse their billing to you. Better yet, ask a competitor of theirs to analyse your billing.
7 Consider leasing out your supervisors to an owner or other construction service buyer. Many of them have no interest in long term employees. Instead, they typically want expertise for one project or phase of a project.
8 More lawyers than ever. You may want to review your attorney's current fee and expense arrangement.
9 Trust the site but verify. Find ways to supervise the site more directly and lessen needed staff. With today's technology, there are ways to accomplish this.
10 Focus on output, not activity – billable/collectable output is best.
11 Slow down to do it right – slow is steady and steady is fast. As a result, rework costs approximately three times the estimated cost.
12 Convert more paperwork to electronic forms.

13 Redesign forms. Template and fill in boilerplate parts.

14 Train site supervisors to recognise and solve problems – train their eyes to see.

15 Do not cut people first; cut tasks and re-engineer. What do clients value and pay for? What method is most efficient to install work regardless of its elegance.

16 Do not expect a quick fix. Keep yourself in business to fight another day. It will not be glamorous again for a while. However, the longer you are in business, the greater the chance that luck will find you.

17 Remember why you love this business. Even though you own and lead the company, you may have to be a part-time project manager or estimator.

18 Ask "what do we need to stop doing" to all your employees. This is a long discussion and prompts beneficial thinking. You will learn something.

19 Ask for a cash price. Maybe even upfront payment price. We are in a cash flow-sensitive industry. Payroll is due on Friday. It is an excellent economy if you have cash. Deep discounts are available for those who can pay. In one sector, we have seen a 9% discount for payment at delivery.

20 Consistent early payment of invoices means you can demand more service (such as same-day quotes, timely credits, and prompt pick up of returned material). This is a more significant gain than the small amount of interest lost.

21 Make sure what supplier pricing column you are in. 1 or 10 is best (for 10 column pricing matrixes) depending on the supplier. Know where, why and negotiate.

22 No colour printers, only colour presentations via a projector.

23 Keep copier lid down so as not to waste toner on edges of copies.

24 Use scanner-to-email to lessen printing, postage and handling on your end of the equation.

25 All personnel must be open to working on project sites and yard settings, it may come to that.

26 Withhold renewals until the last possible second. An unsigned renewal or new contract engenders doubt in the mind of the vendor. It is a leverage point for a one-time concession.

27 Know what drives your economic model, cash-to-cash cycle and motivates clients.

28 Eliminate business and personal combinations (start with top management) – no sacred cows, bulls, or cicadas.

29 Get serious about a preventive maintenance plan (use job site checklists; evaluate total cost, including repair costs).

30 Investigate lease-purchase and upgrade options. Beta testing of a new construction or office equipment can be a windfall.

31 Employ a part-time mechanic. Someone is always looking for this kind of work on a half-week basis.

32 Place working shareholders in the proper salary range. Do not let them be paid above the market for the job they perform. This makes the trip from gross profit to net profit shorter and more efficient. In addition, it keeps the shareholder focused on producing profit (for dividends).

33 Counter or walk-in purchases at the supplier are usually not given the lowest price. This should be addressed. There is no delivery cost, minimal counting errors and product misordering. It is more efficient for the supplier in several ways and thus, you should receive a lower price.

34 Negotiate everything. It is the highest and best use of any professional's time on a per hour basis. Do not get tired of being frugal. One magic phrase is, "*I think I can*

do better someplace else". Also, whenever a vendor apologises for a mistake, ask, *"Can you quantify that?"* meaning giving something of value is a better apology.

35 Make sure employees know the revenue/cost (loss) trade-off. For every mistake or cost, profit pays for it. So, at 4% net profit, 25 times the loss must be generated in turnover to pay for that expense. It is fantastic; once your staff is aware, they will think of many ways to fight the cost battle. That is, if you can direct their thinking to apply the above ideas.

Payment Retention and Its Impact

Retainage of payment is a common practice by construction funders. It holds back a percentage of income (usually ranging from 5% to 10%) due after the project. It is a practice that building and infrastructure owners use to assure financial security against contractor default. In addition, it is to ensure that the contractor promptly remedies any deficiencies in the project.

Retention has evolved over the years into a constant problem for construction companies. It is not unusual to hear stories of retained payments more than a year old. Frustratingly, this is longer than a contractor's warranty period.

According to a Clemson University research study on retainage practices in the construction industry, all contractors' retention releases on a projected range from 30 to 900 days. The average for speciality contractors is 167 days from completion of the project, almost six months (2004). However, from our anecdotal research, this period seems to have lengthened in the years since.

Many contractors would prefer to ignore retention management because they view other management areas as more critical. While some owners and developers of construction projects wisely use retention management as a shield against potential construction and liability problems, others seem to be in no hurry to pay retention as they do not see it as a real problem. Ten percent just does not seem like much money to them. They see it as sapping their energy from pressing problems.

Accomplished leaders in the construction industry believe that they make better day-to-day decisions when they "manage in the present", when they do not focus on old problems or mistakes. Managing retention well means dealing with an old, nagging issue when they could be focused on making profits with today's projects.

Most of the time, managing in the present truly is the wisest course of action. Nevertheless, the area of retention management is the only time that we suggest managing in the past.

It should come as no surprise that speciality contractors who work in the early stages of a project carry more long-term debt than others. Especially civil and structural contractors who pay large material and equipment costs at full invoice and then wait for the owner to release their retention after later phases have been completed. They remain in line with everyone else, only longer.

Many contractors wish that the entire practice of retention be abolished by law. After all, it is reasoned, we already finance the construction with our funds, never getting ahead of the client until the very last payment. However, since no real change is anticipated in our lifetimes, we will have to manage it as best we can.

Let us first say that many owners do not abuse retention. Holding a percentage due back to ensure satisfactory completion of the project makes good sense. That said, retention will be a source of financial hardship if payments are still retained after a project is completed.

As the laws are different from state to state, you may wish to retain a local attorney in extreme cases. These laws change and each situation is unique. Knowing an effective attorney to untie the retention knot is a "must-have" relationship for all contractors.

Payment Retention's Cost

For a construction company, retention costs approximately 20% to 50% of the firm's profit. For contractors whose profit is average, the impact is on the high end of the scale. For highly profitable firms, it is less of a problem.

Retention can be a huge issue for a young firm that may be undercapitalised. Some contractors may find themselves working with mediocre clients over several projects, trying to compensate for that last 10% in lost past payments. Going to the bank to fund this shortfall is the first warning sign that your firm is in trouble.

The cost may be subtle for a well-capitalised contractor as the contractor may self-fund this shortfall without relying on a bank. Self-funding to cover retainage issues is a primary benefit of being financially conservative, which lessens the impact of the problem. Still, even this strategy ties up funds that could be making you profit if invested.

The total impact depends on several complicated factors. However, as a starting place, refer to a sample retention analysis (see Table 5.1).

Month: The months that the project is (or was) built.

Earned Amount: The dollar amount of earned billings against the contract per month.

Billed Amount: The amount billed to the client. It will be less than the earned amount since retention is withheld.

Remitted Amount: The amount paid in the month received.

Direct Cost Expended: Dollars contractor spent building the project. This includes labour, material, equipment, subcontracts, general conditions and other direct costs.

Net Profit Earned Before Tax: Dollars earned after project and overhead expenses are paid before taxes are subtracted.

Cumulative Shortfall Against Cost: The job expenses that are being funded by the contractors which are not covered by the payment by the owner.

Interest Cost of Shortfall (10% APR): The cost of money or the interest charged by your lending institution for the amount borrowed.

Retained Payments Interest Cost: The cost of money or the interest charged by your lending institution for just the retention amount. This is the cost of retention to the construction contractor (see Table 5.1).

Payment Retention Management Strategies

Carefully Engage with One-Off Developers and Project Owners: If you work with people who do not make construction a continuous business, you may want to "pass" or propose eliminating retention. To justify elimination, point out that the owner is cash ahead at any point of a project and other risk management practices that negatively affect the contractor.

Condition Your Bid: Managing retention starts at the bidding stage. A bid proposal may contain language that states your retention policy. You should be the first to propose how it is to be handled. This allows for a frank discussion with the client about this issue. (for more discussion.)

Federal, state, or local projects have no room for negotiation, but keep in mind that some states have outlawed retention on public projects. Private work, on the other hand, has a less standardised approach. As a result, more people are taking this course of action and use a multi-fold approach: (Table 5.8)

1   Enlighted retention managers typically ask for a 50% reduction at the project's midway point. The mid-way point should be calculated by the per cent complete method (versus the estimated time to completion).
2   All punch-out items still on the list three months after project closeout should go to a warranty item with an appropriate value is held back by the client. Three times the estimated cost is a good guideline of the holdback. Retention less than the three times warranty item value is paid.
3   If your philosophy is to charge interest on late payments from the client, use double the prime rate as your interest. It is double because it is fair to charge the cost of money, including a penalty. Of course, some firms use the maximum statutory limit in their state.

1   Many contractors condition their bids on other areas to offset retention's impact to receive better terms and conditions. As an example, they may require payment to be made quickly after the approval of monthly billing. Some create a schedule of cash flow positive values for most of the job and make that payment schedule part of the contract. It is part of the bid package and subject to any pre-contract negotiation.
2   Refuse any contracts that do not explicitly detail how retention is to be handled. At a minimum, redline the contract and then have a face-to-face meeting to negotiate. Again, your professional association or attorney will have standard language that will protect and expedite your retention rights.

No New Bids: Do not indulge clients who do not pay in a timely fashion. It is perfectly fair not to bid more work if payments (retention or otherwise) due have not been made.

Personal Presence: Sitting in someone's office until they pay you is awkward and confrontational, but it works. Let the client know that you will be coming by for your retention before you show up. Tell them in advance that if they are not there, you will wait. Tell them you are serious. Be friendly but firm. If you have patience, this eventually works, although you may have to have lunch delivered to show how serious you are.

Partial Acceptance: Sometimes, the next-best strategy besides reducing retention is to accept parts of the project, such as floors on a high-rise where access can be controlled.

Punch-As-You-Go: We have observed many contractors using a punch-as-you-go method to ensure fewer punch list items remain at the end of the job. This is an internal process of a construction firm that is part of a monthly project meaning. Significant punch out items are listed and made part of an action plan before the next

Table 5.8 A Retention Cost Example

Month	Earned Amount	Billed Amount	Remitted Amount (45 days A/R)	Direct Cost Expended (87%)	Overhead Cost Expended (10%)	Net Profit Earned Before Tax (3%)	Cumulative Shortfall Against Cost	Interest Cost of Shortfall (10% APR)
1	$50,000	$45,000		$43,500	$5,000	$1,500	–$48,500	–$598
2	$75,000	$67,500	$45,000	$65,250	$7,500	$2,250	–$76,250	–$940
3	$125,000	$112,500	$67,500	$108,750	$12,500	$3,750	–$130,000	–$1,603
4	$150,000	$135,000	$112,500	$130,500	$15,000	$4,500	–$163,000	–$2,010
5	$200,000	$180,000	$135,000	$174,000	$20,000	$6,000	–$222,000	–$2,737
6	$150,000	$135,000	$180,000	$130,500	$15,000	$4,500	–$187,500	–$2,312
7	$125,000	$112,500	$135,000	$108,750	$12,500	$3,750	–$173,750	–$2,142
8	$75,000	$67,500	$112,500	$65,250	$7,500	$2,250	–$134,000	–$1,652
9	$50,000	$45,000	$67,500	$43,500	$5,000	$1,500	–$115,000	–$1,418
Retention billed mo. 10		$100,000	$45,000				–$70,000	–$863
11							–$70,000	–$863
12			$100,000				$30,000	$370
Total	$1,000,000	$1,000,000	$1,000,000	$870,000	$100,000	$30,000 Net Profit Before Tax		–$16,767 $13,233

Notes:
- Assume $1,000,000 project, 9-month schedule.
- Assume no overbilling/no under billing.
- Assume labour payroll is weekly and material/equipment A/P counterbalances to average days used above.
- Assume 45-Day Payment Average.
- For each month that final retention payment is delayed, add $863 per month.

monthly meeting. It eliminates the rush of activity at the end of a project to address a long list of items.

Discourage Multiple Punch Lists: Your goal should be to receive only one punch list to enable you to focus on that to completion. Thus, the job will be quicker to complete, triggering the faster release of retention.

Reverse Retention: Ask for an advance of payment on the contract. In construction, all things are negotiable for the craft-savvy and business-minded contractor. Contractors that deliver quality, speed and safety can ask and receive special consideration from clients, such as advance payments. As necessary, they can expect to receive it.

Offer a Slightly Reduced Price: Some construction firms have offered a slightly reduced price to the client in exchange for no retention. The complete calculation of your company's percentage cost is the topic for another article. Furthermore, as clients negotiate hard for you to meet a lower price, this is a worthy negotiating point.

Government officials and others have experimented with 5% retention and even no retention. Nevertheless, a 10% retention policy remains the norm. Couple this with a 4% average net profit for a construction firm, which is evident why it will not change.

The battle of retention management will continue. We cannot predict any change to retention practices. Nevertheless, it is powerful leverage in assuring a fast completion of a project. If we built for our account, we would be foolish not to use it. After all, what better lever is there for ensuring satisfactory completion than holding back triple the net profit?

Managing Financially in a Recession

Many recession-affected contractors move from their sector to others – possibly to other states. That is a rational business move. However, they will not see their market coming back swiftly. There are many other such strategic redirections by construction firms in poor economic times. Your strategic challenge is to anticipate others' moves while planning your own proactively. A yearly review of this part of your plan is advisable. Construction recessions are never a question of "if" but "when".

If a construction firm wastes good economic time by not reinvesting part of its profits and reinventing itself, it may see a sharp reversal of its fortunes. This is sad but fair. The employees will suffer somewhat for the owner's oversight. This business will be less of a competitive force than those who took the fiscally conservative route.

During economic turmoil, make sure you keep track of where you are financially. The stress of a cash flow problem is not the only reason to lay off core employees. In our opinion, another important measure is the gross profit per labour hour (or labour week). Since we are in a variable-cost business, this makes economic sense. Numbers in your normal historical range tell you that the company is in decent financial shape. Therefore, we believe this is an imperative measure.

When the economy cools, it is no secret that keeping your core employees together is wise – both site and office staff. This is your means of production. If you see a bleak outlook and a decision must be made, most people cut back on their personal financial needs, including paying themselves last. This ensures that when the economy improves, they will quickly catch the wave of growth again and, thus, profit. Contrast this with others who, once have contracts in hand, must rehire, or worse, are forced to find new employees, all of whom will need training and make a higher number of mistakes. These firms will be catching the same wave that is, less profitable.

Collect you are Accounts Receivables as much as you dare. Mature businesspeople know that money is a company's lifeblood. Your legal and ethical attempts to collect are reasonable. Your fair pursuit should not be forgotten in the months to come. Any firm with trouble doing this is sure to have difficulty being profitable regardless of the quality of its work. As we discussed at length in the financial management chapter, the only rule that matters are "Do not run out of cash".

Accounts Payable processing may represent one of the most significant potential time savings for office personnel. Reviewing and streamlining the invoice handling and payment process can be a leverage point. If you have a dozen projects, you may be processing hundreds of A/P invoices each month. Using technology and processing take wasted time out. Do not forget, consistent mistakes by a supplier should be grounds for a penalty. Keep the A/P area from distracting focus away from installing work.

Use your financial ability to pay as a strategic weapon. Some firms can pay in a timely fashion regardless of the economy's condition. This is stellar financial management in action. However, in any economy, it should be considered that payment should be made differently. Those suppliers, service providers and others could be divided into three A/P days' groups depending on their service level, efficiency and overall performance. Said differently, all your project partners do not act the same towards to you. So, treat them differently. Do not pay them the same, especially the ones who are a drag on your business. Negotiating based on your strength to pay bills promptly is not a new idea.

6 People

People who are clients and employees make for most of the risk and reward in construction. They enable project progress and affect cost. Communication is the way everyone attracts or dispel others. Knowing more about people is the second step to effectively communicating with them – of course, the first is knowing yourself.

You can never know too much about others. As our business is highly stressful, you need to understand how others may react and how you present truthful information to them. Noted behaviorists Tversky and Kahneman stated, "People are not complicated, but the relationships between people are" (Lewis 2017).

Client communication should be carefully considered. It is a linchpin to critical issues such as problem resolution and timely payment – twin killers of construction companies. Clients will have preferred ways to communicate, such as cell phone, email, fax or letter. Be sensitive to their preferred way. Most value-oriented clients are buying timely delivery. Unfair as it may seem, they do not want to hear about most small, solvable problems. Part of your value is to deal with trivial matters without tying up their time.

A Critical Part of Any Strategic Plan – Low Maintenance People

Construction people are a normal group in the population. They are diverse in culture, politics, religion, education and life experiences. They are our friends (Green 2011). However, contractors desire people who are "low maintenance" (LM).

Questions: What is the great "wild card" in construction contracts? Where can you find an opportunity to have a unique advantage over all of your competitors? Answer: In your people. More specifically, in finding, hiring and retaining LM people in your business.

All professionals (construction or otherwise) have experienced this LM person phenomenon. Since most of us have experienced far too many "high maintenance" (HM) professionals, we know firsthand the time and effort needed to maintain working relationships with them. After experiencing that kind of dynamic, most of us desire LM employees.

We define LM professionals as employees who seek to do an excellent job independently. After initial training, they rely on their supervisor for direction only when starting projects or help in extraordinary circumstances. Otherwise, they are self-starters and do their job well without micromanagement. These are our observations:

DOI: 10.1201/9781003290643-6

- They compete against themselves (not others).
- They manage in the present, forgetting yesterday's disappointment and tomorrow's great event.
- They use less emotion and more objectivity in their work.
- Work is not the most important thing they do in their lives.
- They do not seek trappings of position or wealth. In other words, they do not try to differentiate themselves visibly from others.

There may be many reasons for people have an LM attitude. It does not matter what they are. What matters is that they need less intervention to complete their assigned tasks. Thus, finding, hiring and retaining LM professionals allows management to lead and manage the firm with more time and fewer crises.

To start on the pathway to understanding the LM profile, an analysis instrument is crucial to understanding people's personalities scientifically.

The DISC Personality System

The DISC Profile Methodology is a helpful device to understand behaviour patterns and emotions. This aids contractors to identify employees' strengths and weaknesses so that they can capitalise on strengths and remediate weaknesses for the benefit of the firm and the employee. DISC is also a good tool for evaluating prospective employees.

Several caveats should be kept in mind when analysing DISC information. First, no one should attempt to change their personality. To do so would probably be a waste of effort. It might work for a short period, but the strain would eventually force you to revert to what is natural.

Second, there are neither "superior" nor "poor" profiles. We have found successful and incompetent, happy and miserable, personable and unfriendly people who score within the DISC behaviour profiles. Third, an ingredient that frequently appears in successful people is not one personality type but awareness and accommodation of the strengths and weaknesses in one's personality, whatever the type. If someone was limited to two words to describe this ingredient, it is self-awareness. The DISC process aids in self-awareness and can recommend tactics to accommodate one's personality to win at their job more often. Although specific behavioural patterns adapt more quickly to certain types of jobs, the DISC is handy as one source of information for making staffing decisions. Keep in mind that all kinds of behavioural patterns have proven successful at work.

Third, the information contained here is general and should not be used as the sole criterion for making personnel decisions. For example, the DISC does not measure IQ, experience or talent. Instead, that determination is made by checking references, administering intelligence tests, reviewing background data and trusting your intuition based on successful experience.

The DISC model personality dimensions are:

D – Dominant or Driven Personality
I – Influencing or Inspirational to Others
S – Steady or Systematic about Work
C – Compliant or Conscientious of the Rules

Each individual uses these dimensions to some degree, which may or may not vary depending on the individual's circumstances. The behavioural states the DISC evaluates are perceived expectations of others (and job), pressure and self-perception. Most individuals will use one of the four dimensions more than the others under a state of strain. This dimension is called the "high" or primary characteristic for this explanation.

Highly dominant or driven individuals attempt to shape the environment by overcoming opposition to achieve results. On the positive side, these persons may be quick decision-makers, good problem-solvers, unafraid of uncertainty and new challenges and adept at getting results. On the other hand, they may also be impatient, pushy, easily bored, quick to anger and uninterested in detail.

Bottomline minded persons usually prefer to work in situations with authority, prestige, lack of supervision, various challenging tasks and opportunities for advancement. They often are effective when working with others who are more inclined to calculate risks, are more cautious and more patient with others.

To be more effective, highly dominant individuals should seek challenging assignments that require techniques based on practical experience, make an effort to understand that they need cooperation from others to succeed, identify reasons for their conclusions and pace themselves. Occasionally, they need a shock if it becomes necessary to make them aware of the need for different behaviour.

Highly influencing or inspirational individuals attempt to shape the environment by obtaining the alliance of others to accomplish results. These persons put effort into making a favourable impression, helping and motivating others, entertaining people, generating enthusiasm and making personal contacts.

These inspirational individuals usually prefer to work in situations where they have social recognition, group activities, freedom from controls and details, freedom of expression and favourable working conditions. They are often most effective when working with others who speak out candidly, research facts and details, concentrate on follow-through and deal with things or systems rather than people.

To be more effective, highly influencing individuals should try to manage their time if their dominance or steadiness dimensions are low. This will help them meet deadlines, apprise others on more than people skills, be more objective and be firmer with others.

Highly steady or systematic individuals cooperate with others to carry out tasks. These persons are usually patient, loyal, good listeners and slow to anger. They may dislike noisy arguments and will calm down excited people. They are capable of performing routine or specialised work.

High systematic individuals usually prefer to work in situations with security, slight infringement on personal style, infrequent changes, identification with a group and sincere appreciation for their contribution. They are often most effective when working with others who delegate, react quickly to unexpected events, work on more than one thing, apply pressure and work flexibly.

To be more effective, these steady individuals should pursue multiple strategies. Such as condition themselves for change before it occurs, seeking information on how their efforts contribute to the overall organisation, collaborating with co-workers whose competence they have confidence, asking for guidelines to accomplish tasks and seeking out those who encourage creativity.

Highly compliant or conscientious individuals attempt to promote quality in products or services under whatever circumstances exist. These persons usually follow directives and standards, concentrate on details, comply with authority, criticise, check for accuracy and behave diplomatically or formally with others.

Highly conscientious individuals usually prefer to work in situations where others call attention to the high compliance individual's accomplishments, security and standard procedures, identification with a group and the absence of frequent or abrupt changes that might compromise quality. Thus, they are often most effective when working with others who delegate, seek more authority, make quick decisions, compromise and use policies as guidelines instead of rigid rules.

To be more effective, high compliance individuals should seek out precision work, tolerate some conflict, appraise people on more than their accomplishments, ask for guidelines and objectives and set deadlines for planning and deliberating.

As a rule of thumb, most people possess two personality traits above the "mid-line" or average emotion. They have two below. If three are above or below, this may indicate a change of circumstance – good or bad. If all four personality traits are close to the mid-line, then the same is true again. People in transition emotionally (some-times due to a new job or traumatic event) are in a temporary flux mentally. After some time, their true personality will surface.

DISC does not classify according to creed, colour, race, national origin or another characteristic. In the parlance of Human Resource Management, it meets Australian legal requirements 4/5ths rule. This is valuable in today's employment climate. For a more detailed understanding, see Table 6.1.

Generations in Construction

That feeling of alienation felt by older workers toward the younger generation is not new. This is part of the very definition of a generation. A new generation defines itself by creating a new "biography". A new generation crafts its identity in part by attempting to correct the mistakes of the middle-aged generation. To understand your generation and others, we need first to examine each. There are currently four generations in the workforce:

• The Silent Generation (born 1925–1942). This generation came of age during a

Table 6.1 Work/Personal Interaction Matrix

STYLES	Excellent 1 2	Good 3 4	Fair 5 6	Poor 7 8
D - D		P	W	
D - I		P	W	
D - S	W		P	
D - C			W	P
I - I	P			W
I - S	W		P	
I - C		W		P
S - S	P	W		
S - C	P	W		
C - C	P	W		

Source: Stevens Construction Institute – Archived Material.
P = Personal Relations; W = When at Work.

peace span, affluence then disrupted by economic depression, war, and disloca-
tion. The youngest will be well into their senior years and are likely owners or
senior workers near retirement in construction companies.

- The Baby Boomers (born 1943–1960). This post-war generation came of age in a
turbulent time including The Vietnam War and social movements. The older
generation members are retired; the younger members are currently in senior
management and leadership roles and do not interact as frequently with the
Millennials as the next generation, Generation X.
- Generation X (born 1961–1981). "GenX-ers" is the bulk of the older workforce in
construction; they fill most middle and many senior management roles. They are
generally the direct supervisors of Millennials.
- Millennials (born 1982–2002). In the Millennials' case, they look to the self-
centeredness of the X-ers and seek a new identity that is more connected, more
civic-minded and more community-oriented. Although we may have heard them
referred to as "Generation Y", this is not an appropriate moniker (Garrett, 2011)
https://apps.dtic.mil/sti/citations/ADA559872. It is tempting to see one generation as
an extension of our own. This is not the case; they are not simply the following letter
in the alphabet; they are not an extension of Gen-X. They are something else entirely.

It is important to note that each generation is raised by the previous one. What is not
apparent is how parents express life's experiences to their children. The previous
generations have shaped attitudes and actions. The resulting work ethic and professional perspective matter greatly to employers and their managers. Again, we believe it
is logical (a predictable result) that the current generation is who they are.

In summary, working with people in acceptable ways to all generations is what we
are suggesting. The largest group of available craft and management talent readily
resides in your era. There is no other group that has the population, proximity and
language skills. This logic cannot be denied. It applies not only to your internal staff
but those outside parties you engage. It is obvious; you engage this generation each
day. How to engage them better is the million-dollar question.

The possible directions and actions shared are conclusions drawn from research.
Consider them as potential solutions. An artful approach is needed when applying
each to your unique workplace. Take some time to consider what, how and at what
pace to implement each.

Recruiting, on-boarding, managing and motivating is still critical for all generations; however, each is slightly different in its turn-ons and turn-offs. Nevertheless, it is
an essential skill set and challenge for many of us. The good news is that some contractors, including their youngest staff, do not want to address this change and
therefore, a recruiting edge goes to those who do.

If we become frustrated, it would be wise to remember that all generations complain
about each other. Your generation will be complaining as forcefully about the newest
age as the oldest generation has about them

Growing Our Workforce Generatively

All firms have a culture, whether intended or accidental. When that culture is not
satisfactory and a firm grows, the addition of unknown people can amplify the
problem. These employees are needed to take advantage of a construction

company's growth, yet they undermine its goals by being out of sync with its culture. A group of unknown people can change how the company responds to adverse circumstances, making this response unpredictable. In some cases, unfamiliar people have not been exposed to the firm's culture long enough to adopt the desired behaviours. In other cases, the firm does not have a sufficiently strong culture that incentivises new people to embrace it. Problems can also arise when new people attempt to change the culture.

In short, if growth is in your company's vision, you should be paying attention to culture, so your current profit percentage is maintained or improved as your turnover increases. To this point, author Greg Satell suggests that great companies do not adapt; they prepare. That is our experience. We have observed that executives in well-managed firms have a vision for the present and some years in the future. They plan for today's reality and know tomorrow is on its way. They have intuition and sensitivity for industry events. They project the meaning of any significant event into the future.

As an example of good leadership and management in construction, the quality should not be a source of complaints if you furnish small tools or equipment for your employees' use. However, as good craftspeople are hard to find and labour efficiency is a crucial success factor, scrimping in this area can lead to adverse project outcomes and hard feelings among your means of production – your site forces.

In our view, the progression of Management Models is primarily driven by a risk focus that is often part of increased employment. The owner-operator who constructs work can be by himself or leading 100 employees. It is a choice to be in control of most things or to delegate.

The positive progression of an Operational Mentality is driven mainly by a *people focus*. It is a learned behaviour created and continued by many iterations of exposure to the company's norms. Cyclical events (such as gatherings, awards and newsletters.) do not set a strong culture in an employee's mind. From our observation, culture results from consistent company actions, especially people who are supervisors. For many construction people, the company is their immediate supervisor. In the case of a poor manager, many people have whispered something akin to "Why does the company continue to employ this person?".

Moving an organisation's culture to the desired place is a long journey. However, it is subject to interpretation by all employees, with the most company experience ones the most crucial. We suggest that there are large and lasting benefits to be gained for any firm that shifts its Operational Mentality to one where people are enabled and prompted to engage in self-and company improvement.

Foundational Components of Culture

We observe three main drivers of a company's culture in construction contracting firms: 1) People, 2) Processes 3) Company Infrastructure. This section focuses on people.

People: Hiring the right people is the first step. Culture is maintained and deepened by bringing people into the firm who are mentally and emotionally aligned with the vision and mission of the firm. In contrast, if we hire for technical performance only, our culture will be a mishmash depending on which crew we are discussing. In times of growth, it is tempting to hire whoever is available. However, growing too fast can be counterproductive. Think of the untested subcontractor who has lost money. Now imagine that person leading an internal team. Getting good people is critical to growth.

Hiring productive, thoughtful employees is not enough, however. From the employee's first day, an employee must be given the proper examples and incentives to adapt and evolve their thinking into the company culture. Simultaneously, we do not want the culture to be bureaucratic or pathological, so it is essential to allow new people to shape and improve it. Culture is complex! It can start with an onboarding process for an employee and be strengthened through culture-building activities for all employees – like mentoring programs.

It is also evident that middle managers are not aware that culture is a guide and motivator to unknowing employees. Additionally, they may not understand that they are the immediate representative of the company to their subordinates and as such, they set a culture. So, they need to set it for their colleagues to witness.

Lastly, a company may conclude an employment relationship with an exit interview. There are two reasons for this last process:

1   A learning process for the company. Each company can learn from soon-to-be ex-employees the strengths and weaknesses of its operations, including its culture.
2   An "alumni" practice for future business possibilities. Keeping a good relationship with a talented former employee is a good cultural and business move. The company shows its continuing care for people and keeps the door open for their return. Additionally, industry people talk constantly. This last positive "goodbye" is sent across interpersonal networks lifting a firm's image.

The Great Conflict: Office Versus Site

Who started the great conflict between the site and the office? Why do some organisations have site processes that work nicely with their office counterparts and others have a mild war between them? Is it normal and what can be done about it?

Being mindful of what Stephen Covey stated in his book, *the 7 Habits of Highly Effective People,* that "Begin with the end in mind" is critical. In other words, keeping a worthy goal primary drives good behaviours and eliminates bad ones.

The "end in mind" or goal for many construction people is to complete work well to proceed on other work. Projects that satisfy clients provide for a secure construction business. Well-built projects lead to other opportunities that grow companies and careers. Again, as long as the goal is central to a person or organisation's thinking, the work will be constructed well and on time.

All superior construction firms focus on four outcomes:

1   Safe projects.
2   Quality work.
3   On-time completion.
4   Competitive cost to the client.

The processes and people who produce these ensure their company has a certain future. Conversely, any action or method that is contrary to these outcomes makes for uncertainty. Therefore, these outcomes must be central to any organisation and its professional thinking.

All formally organised contractors have site and office operations. Each executes critical functions in the ongoing life of a construction firm. However, they have conflict historically. Whether due to personality differences or process requirements, they do not get along in some ways. It may be because of style, demands or just perceptions, but it can be tragic to construction firms. Any internal conflict is controllable; however, left unaddressed threatens the turnover dollar with slower turns and higher costs.

The site and the office are separate but equal. Critical in their own right and cannot be eliminated. The functions they serve have been around since people have built for a profit.

Office – ensures that a contractor gets the next project, builds it profitably and gets paid for it. To do this, they need to track everything. Therefore, paperwork is a legal and practical requirement.

1   Style – Formal. Written communication is, many times, desired. If it is not written down, it did not happen. Wants detail of all processes and events documented. The building process is ongoing and is high pressure, but the office needs information now.
2   Background – More documented education and more formal training ("certified smart").

Site – makes certain to install work well. All the project parties are managed or influenced by the site staff. Site managers and other project supervisors make sure the company's labour, materials, equipment and other project expenses are coordinated on the job.

•   Style – Informal. Verbal communication is sometimes desired. Reading reams of paper is not preferred. Wants labour, material and equipment resources gathered days before they need them, so can confidently schedule work. Can work out small details later if most everything is in place. Clear and up to date information about pending variations, requests for information and material delivery dates is desired. They also want preferred crews on the job consistently.
•   Background – less documented education and more informal training ("college of hard knocks").

The site and office in many companies have conflicting styles of managing projects but typical roles – make positive things happen on the job and for the construction firm. Unfortunately, the kinds (formal or informal, detailed oriented or not) just cloud the communication and, therefore, hurt the project's successful completion.

However, as a quick antidote to a poisonous site and office relationship, more communication and exposure to each other (not less) will make each side work better with the other. Said differently, people face to face with each other act more civilly than over the phone.

In those companies where each of these parties work well together, there is superior communication and therefore, fewer project problems. Each side "gets it". Any odd personalities and unusual work habits are tolerated. All members realise that the "main thing is the main thing" – that is, the company's success.

In profitable firms, it has been observed that all members understand the business cycle of construction contracting:

* Acquiring work.
* Installing work.
* Getting paid for that work.
* Tracking all the above.

Superior firms know that the faster the cycle than your competitor, the better the competitive edge. Each person cannot do everything to speed it up but does several things to keep that cycle fast.

It is evident in all studies that people want to do an excellent job on the site and office. Rarely would someone say, "*I want to go to work to be demoted, lose the respect of others and take a pay cut*".

So, some in our industry may have to follow a client's advice, "be a bull chewing barb wire". Meaning smile, whatever conflict may come your way. In other words, keep the conversation going, moving forward and be pleasant.

It is essential to add, skilled and active professionals have a certain future. Experienced and proven construction people have more opportunities than others, no matter the economy. In contrast, any construction firm does not have a certain future in our risk-filled industry, no matter the name brand or the depth of funding. Both are not unlimited.

So, the big picture for anyone's career is to be a team player and do a great job. Working well with your opposite, whether on-site or in the office, is the point. This is unemployment insurance.

So, look around and be observant of people. As a friend of mine said, "Be an anthropologist". You might be surprised how poorly people work together. It cannot be tolerated. The good news is that improving it only raises your firm's current performance.

In our 4% average net profit before tax industry, all construction firms cannot allow anything but full cooperation in these competitive times.

Are Hiring Gritty People the Best Fit for Construction?

Angela Duckworth's insightful book, "Grit: The Power of Passion and Perseverance" asserts that those with a steady and determined focus outperform others. She suggests the term "Grit" identifies the overall character of those who have passion and perseverance. Her team's extensive scientific research produced compelling evidence and is the basis for a 12-point assessment measuring this valuable characteristic. This section explores the background and applicability of this trait in the construction industry.

Duckworth asserts that this personality characteristic raises the probability of significant accomplishment. In their 2016 book, she describes many assorted Gritty people, including those who make it through "Beast Barracks" and eventually graduate from the United States Military Academy at West Point, New York.

When we read Duckworth's book, we immediately thought, "Most people we know in construction are Gritty". That is, we know many construction professionals who, we believe, meet Duckworth's standards. But, then, we thought about it some more. We spent much time talking about our colleagues in construction – people who embody passion and perseverance, sustained over time, toward a focused vision for themselves and others. We wanted to highlight Grit Paragons, who exemplify what it means to have passion and perseverance.

We have interviewed and worked with hundreds of construction company owners and managers, from CEOs to Forepersons. Our first interview was in 1994 and continues currently. This section describes four of the Grittiest people we know in the construction world – a distinction heightened by the industry's general hardiness.

The Logic of Grit

Passion and Perseverance are the two characteristics that combine to form a Gritty personality. They are two parts of a whole – together, they form an attitude. When a Gritty person pursues a heartful vocation, they do it for a long time. The focus is not a short-term obsession or infatuation.

According to the research, Grit is a small part of our nature (genetics) but emerges primarily from our nurture (experience). So, it is a product of one's existence. Duckworth concludes that Grit grows during one's life. She suggests adverse events and a person's determined reaction to them are shared among the Grit Paragons.

Thus, this insight gives us a basis for how we might look for those who have it. It follows that questions about a person's life journey and how they view it become more important in a hiring interview process.

As an example, Duckworth describes how Seattle Seahawks football coach Pete Carroll was sacked from his first NFL head coaching job. The experience was challenging, but he reacted in a positive and determined way. Pete knew he had to find answers for being a better team leader to win another head coaching position. Carroll's determination showed as he searched and found the answer in a "Life's Philosophy". It suggests that one must have an overarching goal-one that matters to the person. It is the philosophy that drives a person's actions. Carroll states it this way, "A clear, well-defined philosophy gives you the guidelines and boundaries that keep you on track". It is a core belief.

Carroll's Life's philosophy is "Do things better than they ever have been done before". This is notable in that it denotes a search for a better way. Furthermore, it implies he is on a journey with the players. Since Gritty people persevere, a life's philosophy, however, labelled, must be in place in each of them.

Given the complexity and ambiguity of the construction process, we contend that Grit is the consistent ingredient of successful hires and long-term employees. They must have passion and perseverance to keep working in an industry that can be dirty, thankless and draining. The site supervisor who will hang a door when the glamorous work dries up, or the project manager who will estimate and answer phones when the wrong person has the flu on bid day – the kinds of people we want in our organisations.

However, we must state a caution. Research suggests that staying when losing consistently can be costly. There is a value to knowing when to quit even though the act of quitting is objectionable. Highly gritty individuals are well served to modify and change the direction of their efforts when it is clear that to continue is costly (Lucas et al. 2015).

Our Grit Honour Roll

We contend that each person who has been in the construction contracting industry for ten years or longer has high levels of Grit. Their passion and perseverance are evident to us. They have seen a cyclical downturn or two and have been employed continuously by a construction firm. Since they have been constantly employed, each must be above-average in productivity and dependability. They have a body of work that is consistently positive in the functional areas of a construction firm, such as estimating, project management, site supervision, finance or human resources. They have specific skills that are deeply ingrained through discipline and effort.

As you can guess, we know many people in the Construction Industry. However, there are a few who stand out for their extraordinary Grit. They show a positive spirit while overcoming difficult circumstances. Here are their stories:

Lisa worked as a division president for a large mechanical firm. She led a company of over 100 employees who were mostly male. This is a significant challenge for any woman. Furthermore, she was a single mother for part of her career, including covering her presidency. As a result, her career progression was from entry-level.

Two events stand out. The first was the case of one employee who unduly enriched himself through a payroll error. He notified no one of the mistakes for over six months. Once discovered, she dismissed this 20-year veteran of the firm. He worked in the financial area, so Lisa and others had to keep this area operating after the dismissal. It was miserable discovering the status of all financial matters (fighting ambiguity) while transacting and tracking many dollars. Lisa kept this area functioning with dozens of late nights, but she did see it through. The second event was her recommendation to combine one operating division with another, although it would mean the end of her employment. The respect she earned was the right kind from the Grit she showed. We have always known Lisa to have the life's philosophy of "doing the right thing".

Bill's company went through the 2008–2009 recession, experiencing it as all others did: less available work, more competition, cutting headcount, watching each job closely, collecting what was due to him and being more of an operator than an owner. Yet, we know he paid himself last to ensure the firm's credit rating stayed strong and kept any financial commitments made to employees and others through our work with Bill. Also, this allowed him to keep core team members intact and paid their compensation through the tough months. Furthermore, this prepared the company for the eventual economic recovery with extra working capital.

During the recession, he was willing to do complicated installations too small for other contractors – not making much profit – but demonstrating that he was ready to work hard for clients and deliver high-quality work without a large paycheck. Site supervisors were Gritty too. But, they did this extra tedious work with a positive spirit. Of course, today, those same clients have included him in their growth plans and these managers also benefitted in their company career. Also, one could say the firm is well situated for the next (inevitable) recession. Bill's life's philosophy is to look after his work "family".

Harry works for a medium-size main contractor. The firm performs both public and private work. Harry was born with Muscular Dystrophy (MD). We had been engaged by this company on and off for some time. We have kept up with Harry over the years. His intensity in managing projects and estimating work is significant. When the company determined that certain practices were "Best", Harry was the first to learn and implement them. When he worked as a project manager, his disability became so cumbersome that he struggled to board a plane, so he drove himself to distant job sites. When he could no longer travel, he went back to estimating.

We calculated the company's gross profit per person week for each team, project and client on one engagement. Harry's team (he was the project manager) had the highest we have ever computed. If you know Harry, you are not surprised. We often repeat this story to others to honour him and his Grit. His life's philosophy, he says, is to be an example for his children. We smile because we think he is also an example for many others.

Ahmed is an estimator with a construction firm in the Midwest. He is of the millennial generation. Before we arrived, he persisted five years in an environment where he was criticised for his work and not supported. Ahmed did not seek recognition for his successes or defend himself when the site asserted issues with the estimate with a personality type of an estimator. Instead, he took the criticism on board and continued to improve estimates. He never thought of quitting and continued to strive.

We were engaged by the firm and pointed out things to the owner, including 1) Ahmed's work was consistently good 2) estimators have a short time to estimate and bid while the site has a longer time to review and find errors and 3) the company support of him was less than we see in most firms. The owner agreed with this, so; we went to work to help clarify and direct the firm's work acquisition process. The firm changed to "Best" practices such as Bid Captain structure, Bid Modelling and personnel was assigned specific tasks to help keep Tim's focus on the major risk/reward areas of work acquisition. As of our last conversation, they had won four projects in a row with a good margin – Ahmed led the effort. The owner said the "vibe" of the company was different.

Ahmed's Grit was evident to us. His value was easily identifiable; he just needed clarity and a process that everyone would follow. His life's philosophy is that a career path is less critical than contribution.

Is the Industry Distracted Unduly by Education?

Formal education's importance in construction is apparent. The industry needs those who have been exposed to new ideas and can think flexibly. We find college graduates are more aware of a project's ambiguity than others.

What type of people make for a high performing construction firm? Is it the formally educated? We have observed that formal education has carried many professionals a limited distance in their construction careers. In our experience, the Grittiest get the most out of their formal education but may not be reflected in the school's medal. These kinds of people have a high amount of passion and perseverance to manage through most project difficulties. They possess the "stick-to-itiveness" to solve problems and complete projects to the client's satisfaction. High performing firms have the Grittiest people regardless of education level.

Building shelter and infrastructure is fraught with ambiguity. Construction projects are custom-built in a unique place installed by temporary crews who are mainly unknown and nomadic. Plans and specifications are less complete than 20 years ago. The Construction Industry is not strictly organised with a standardised product. In each project, cost and time demands are flexible depending on the client. Gritty people fight through the ambiguity of our industry.

The Metrics of Grit

We have heard executives use different labels to describe the above-average employee, but they are explaining Grit in our minds. If the highly valuable people to a construction firm have Grit, why shouldn't contractors test for it in candidates for employment? Since there is an assessment for Grit (see below), there is less of an excuse not to.

Grit is a combination of passion and perseverance. These two components may be relabelled as effort. Duckworth asserts that the following components are active when people achieve.

- Talent × Effort = Skill
- Skill × Effort = Achievement

Grit's obvious value to construction companies is assumed. We have heard from many clients and their management of its applicability. However, what is the relative importance of other qualities? As you see in the two equations above, Duckworth's comparable measure below makes sense:

- Talent counts once and effort counts twice

From their research, Grit is a concept that can be measured. Dr Duckworth has extended her research and created a tool.

Using the 12-point Grit Scale

Duckworth has created an assessment for Grit. She has field-tested it in several situations. In one instance, the author was given access to incoming students to the United States Military Academy, also known as West Point and tested each before "Beast Barracks" – a highly stressful but necessary exercise in leadership. From the results of their 12-point scale, she found that this assessment accurately predicted those who would fail or graduate. Thus, the evaluation works – it measures what it says it measures. However, in applying good concepts to reality, care must be taken.

First, we ask a question. Could a baseline be sought from your existing personnel as a guide for the minimum acceptable Grit score? If you believe this helpful, then those best at what they do could be a good starting point. In cases of a new job function with no best person, Duckworth provides a relative measure of Grit against the general population.

Next, we debated whether a job seeker should be asked to respond via writing to the Grit questionnaire. We do not think it is this easy, unfortunately. Job seekers are motivated to show their best selves. Therefore, we do not believe that filling out a paper questionnaire will lead to truthful answers.

Both of us sense that a better gauge could be achieved by using it not as a written questionnaire but as:

1   The basis for verbal interview questions, mixing up the order and phrasing (active or passive). Also, we think it is essential to ask for an example after each question.
2   The basis for questions of the candidate's references and again, seeking examples for each question.

Of course, this information would be in addition to all other pertinent background data (such as criminal background, work history and skills training) needed to make an appropriate hire.

In conclusion, Duckworth's book furnishes a framework and explanation of Grit. Construction firms can now better identify suitable employment candidates. Also, she created an assessment so, this trait is measurable. As "Best" practice, contractors can now understand an applicant's level to make better hires.

Although some of it is innate to a person, Grit is grown chiefly from within from their early years. Of course, life's reverses and devastations are not to be celebrated but can make each of us stronger mentally and more insightful. For the "Grit Paragons" in the book, adverse events have made them more determined and resolved to overcome those obstacles in their lives.

The author's study of graduates and washouts at West Point in the United States is one set of evidence of Grit's predictability in achievement. However, several others are presented that support the overall assertion of its importance.

From our conversations with contractors, above-average Grit is a critical component of great managers and executives. We have observed the same in companies – accomplished construction professionals possess high levels. Over decades, those who have constructed have passion and perseverance for our harsh industry. From their personal and professional experience, they have learned which purposeful ways to act to achieve meaningful things in their lives. With higher Grit levels in new hires, we both predict contractors will see better new employee performance. In all, we believe Gritty people are a leverage point for construction companies to grow profitably and deliver consistent value to clients.

Learning and Forgetting

One of our assertions is the erosion of efficiency because of construction's higher new hire and separation rates. One of the stark effects of its relatively high hire and separation rates is workers' adverse learning and forgetting.

It is commonly accepted that the lack of exposure to company-specific knowledge (learning) and not remembering previously learned knowledge (forgetting) significantly affects construction efficiency. That said, the turnover of construction personnel – new hire and separation, diminishes overall learning and increases forgetting.

As noted by several researchers in various industries (some since the 1800s) show that discontinuing a task causes the person executing that task to forget a percentage of the complete knowledge needed to perform it again.

Construction companies' practices are not standardised across the industry. There is no "one way". main contractors lead project work which many times is executed by subcontractors. Their personnel executes these subcontractors' project work with overall direction provided by the main contractor.

This phenomenon becomes problematic when hiring new project managers and site supervisors due to winning a large project. The breadth of scope may be much more significant and riskier than a journeyperson. What they do and say ripples across multiple people and directs those people's actions.

We have observed that new hires are never as valuable or efficient as those who are continuously employed. Multiple reminders and exposures enhance learning to practice.

Rehires, previously employed, separated then, hired again, never work with the same enthusiasm for their company. Indeed, they are vigilant about a possible RIF, constantly calling those they know about contingent employment.

Hiring for current growth and not for long term sustained growth has its problems. Some may say, "That is the way the industry operates". We understand, but it does not follow that an individual company must operate.

Those that pursue a consistent and sustainable turnover have the behaviours of an efficient construction firm.

The Compelling Case for Consistent and Steady Employment

There is a compelling case for consistent and steady work volume versus large and discontinuous construction workflow. Here are five reasons we have observed in our work with construction contractors:

- Variable cost structure means that a construction firm does not have to grow or be large. Variable cost shows that most of the cost is in the production of the product or service and not in the organisation.
- Craft Dependent. People who master and execute with great skill. They plan for productivity and inspect for quality.
- Multiple Custom Inputs but One Specified Output. There are no long production lines in the social impacts. End of employment and its resulting paychecks stress anyone. The employee and their family feel the adverse ripple effects.
- Variable Cost Businesses are one-off and craft-like. They do not have a large machine that is programmable to specifications to produce a single product for the customer. A 4% net profit business such as construction

A less problematic example of discontinuity is the crew that is pulled off a project unnecessarily due to a nervous client who wants more workers. This demand is previously unscheduled, unplanned or uncommunicated to the contractor.

Another observation is that increasing new hires and separations diluted management's effectiveness. When hiring many new people, time is spent repeating instructions, fixing problems, and figuring out where a person can "win". In the case of employment separations, it is an emotional event for some employees. As a result, additional hours are sometimes needed to manage the person and the paperwork of termination.

Setting Project Team Member's Working Relationship and Culture

Industry-wide, there is a lack of a standardised project process. The specifics differ with each main contractor. This represents a learning curve. They lead the subcontractors that install the work. Subcontractors work for main contractors every year, each having its own set of practices. It is little wonder that productivity has not increased over the last 20 years. Coupled with the nomadic nature of frontline management, craft and operator, it is clear to us.

The entire project group and office personnel may not have building work do not have a common framework and philosophical guideline to follow. The Project Partnering furnished that but appears to be little practised on projects today.

The attitude in some sectors is that the "trick" of partnering is once understood need not be repeated. The work involved is not a value. The transformation is that stakeholders create outputs such as a project hierarchy, charter, problem-solving, expectations and action plan specific to an individual project. There is no standard content outputted. It is unique to the project. The outputs are common but the answers are specific to the project.

People seem to become cynical when they made redundant multiple times due to no fault of their own. We find that they learn to "game" of the system of construction. Also after multiple changes, their loyalty and enthusiasm is not channelled at their next employer.

In most project settings, the main contractor, subcontractor and supplier employees did not work on the previous project together.

Sophisticated project owners, funders and leaseholders may be helped by supporting improvement efforts by their contractor-clients to overcome this barrier. In addition, consistently employed contractor personnel are more predictable in their behaviours and outcomes due to a consistent immersion in a construction company's practices.

Improving an organisation's productivity gives it the best chance for continued success. In construction, it appears to be the first concern of most contractors. The business rewards speed and efficiency with profit (not size or brand name).

We assert that reducing the variation of the number of people hired and separated in a construction firm's employ helps improve its productivity. This might be as simple as limiting turnover and backlog a firm earns in any one month.

To crystallise, management actions to consider – when targeting employment discontinuity as a root cause of low organisational productivity:

- Determine a level of income to keep new hires, re-hires and separations minimised. So, a plan for work acquisition personnel is needed when backlog turnover (with a safety factor) is attained. Indeed, some members of this group should be assigned to projects as site staff or project management. Some might even perform particular tasks such as internal and external data analysis.
- · Keep the geographic reach of projects targeted the same. As construction people know, site staff will sometimes quit as family and local friends are a more critical part of their lives.
- Keep the average contract size of projects the same. This will result in familiar demand for resources (not an unfamiliar and severe one).
- Institute a formalised planning period with project-related deliverables. Before starting any project, project owners, funders, and main contractors should consider a formalised planning period. As part of this upfront activity, project owners should forward part of the contract amount to pay (and encourage) this. The process accomplishes two objectives:

  - To set the project team culture and relationships. This takes time in any event, but we believe it is less effective while the ramp-up of activity at the start of a project.
  - To enhance planning. Outcomes produced such as a Rummler-Brache execution plan, CPM schedule, project letter of execution, a problem-solving process and BIM model will prompt all project stakeholders to understand and discuss project details and formulate a path forward before mobilisation.

- Ultimately, a group of contractors (a main and their subcontractors) with an established project culture will build better safety, quality, schedule and cost with fewer claims for time extensions and cost reimbursements. This benefits the ultimate user, the project owner and those who fund the project.

Those seeking to master and ultimately succeed in their construction projects must address this industry's peculiar barriers, such as high employee turnover. Not overcoming specific barriers' effects such as this dampen adoption of production methods such as Lean. To ignore them is to invite suboptimal results. To address them thoughtfully is the start of being faster, better and smarter. This is one in that innovators behave differently as they disrupt and improve the industry for themselves and others.

The Largest Input: Organisational Culture

The core idea of this section is that construction firms often fail to commit to a management model and company culture. That failure to commit generates unnecessary difficulties in leading and managing the business. We have witnessed the results, including suboptimal outcomes. Executives are responsible for determining the best management model and setting the company culture in all respects. This is a daunting task, but one that has a high payoff.

Many companies in the last two years are experiencing turnover growth. For this growth to occur, additional people and assets have been added. Furthermore, this growth has exposed undefined and poorly communication practices leaving each person to complete tasks as see fit. All these inputs can cause suboptimal performance in a time when top quartile profits are attainable. All executives realise the problem; however, each may not have a framework to assess and improve their internal business environment. This section outlines one possible approach to addressing this issue.

Foundational Components

We observe three main culture components in construction contracting firms: 1) People, 2) Processes 3) Company Infrastructure.

People: Hiring people is the first step. Those who bring into the firm who are mentally and emotionally aligned with the vision and mission of the firm is how culture is kept and deepened. In contrast, if we hire for technical performance only, our culture will be a mishmash depending on which crew we are discussing. In times of economic growth, it is tempting to hire whoever is available and thus build more work.

From the employee's first day, a culture-building process can be undertaken. It can include an onboarding process of the first 90 days, then evolve into a mentoring program. Last, a company may conclude an employment relationship with an exit interview. There is a twofold reason for this last process

1   A learning process for the company. Each company can learn from soon-to-be ex-employees the strengths and weaknesses of its operations, including its culture.
2   An "alumni" practice for future business possibilities. Keeping a good relationship with a talented former employee is a good cultural and business move. The company shows its continuing care for people and keeps the door open for their

return. Additionally, industry people talk constantly. This last cheerful "goodbye" is sent across interpersonal networks lifting a firm's image.

Processes: Processes are ways in which your employees work with each other. These processes can be viewed as company instructions for success. As everyone knows, no one person constructs alone. People are part of the extended transaction chains in construction contracting. If unclearly defined, uncommunicated or poorly executed, those processes slow outcomes as the transactions' chains are completed. From this poor process instruction and execution, a culture of conflict and frustration emerges many times.

Of course, companies who clearly define, communicate, document and then, monitor the completion of tasks are rewarded with better than average outcomes. In many cases, they reach the top 25% of peers in performance, especially in return on investment.

Company Infrastructure: The quality and practicality of company assets such as offices, equipment, technology and tools affect project outcomes. It sends a signal when a firm keeps its less-than-new assets well-kept and maintained. However, not having a conscientious program of repair and upkeep can sour front-line managers and others about value and cost. In addition, inadequate equipment, tools, software and other assets make people perform unnecessary workarounds. Eventually, some choose not to engage closely with their duties.

These areas will have to be improved if outcomes are to be better.

Now, let us explore the characteristics of types of management models and, separately, organisational cultures. Then, we will explore how they interact and produce unique environments.

Management Models Summarised

We have witnessed these common management models in our industry.

1   Owner-Operator
2   Family
3   Team
4   Bureaucracy

These help answer the question, "How do contracting firms approach the business of construction?

An *owner-operator management model* is one in which one individual – the owner – makes all the significant decisions. They operate the firm with minimal input from others. It is often the force of the owner-operator's personality that drives the success or failure of the firm. This person is the most significant constraint to improving operations and growing revenue. Of course, others help, such as acquiring work, constructing work and tracking the processes. Still, the owner is the person who determines largely by himself what the company bids, the adoption of a software package or the answer to a site problem. Each of us can be too conservative, too lax, or too emotional on any one day. There is no balancing by another person(s) of this human tendency. This is a hazard to the firm's health since the industry is the second riskiest. We have observed that this culture is most evident in the smallest firms.

We have seen many firms with an Owner-Operator model. Employees tend to focus on the firm's leader more and less on their job duties. It is rational. If raises, promotions and perks are given arbitrarily (not systematically) by the owner(s), then employees will always think of the leader's reaction first. This is rational. Without a check and balance system, each company member can only assure their own prosperity by satisfying the leader's perception of activities such as good decision-making or other's faults in a poor outcome. This politicisation can encourage the most talented to seek employment elsewhere in another, more systematically leading and managing a company.

A *family management model* tends to be evident in smaller firms that have grown a little too large for an owner to oversee all operations directly. Sometimes relatives are many of the employees. Other times, the owner has trusted employees who are "looked after" (in professional and personal ways). For some employees, this is a significant value. Whether they come from a split family or not, a sense of togetherness appeals to many. In a family model, the chain of command is less critical since the owners tend to talk to any employee regardless of rank or tenure. Consensus building, although quietly practised, keeps relations strong inside the company. Bonuses tend to be discretionary.

From our experience, family management has advantages as well as disadvantages. One of the clear advantages is the want of many workers for a family atmosphere. Many workers have suffered a parental divorce or one of their own. A sense of interconnectedness answers this need. However, one of the disadvantages is that the most ambitious employees will not be rewarded for their hard work and savvy. This demotivates and may lead to a person's employment exit.

A *team management model* models itself after sports dynamics. The team consists of some stars, a few tier two and less talented players. However, each person supports winning. Winning is defined as a completed project within the contract requirements with a satisfied owner. No one person does everything, but each does something(s) to improve outcomes. Getting the potential out of each employee is done. Bonuses tend to be determined by measurement and reward both project and corporate success.

Sports backgrounds abound in construction worker populations. The dynamics and cultural mores are well known in our industry. The team management model is evident in parts of each construction firm we know. Although a company is rarely totally immersed, it exhibits elements of the other management models.

As in sports, most teams win by having a few stars with a solid supporting cast. So, talented people must be part of the team. The others who are roles players know what to do and when to do it. Jealousy is not part of a good team. A superior coach who organises the group and keeps the environment collaborative as well as disciplined is essential. There can be no two sets of rules in a tight-knit group.

Bureaucracies, as the perception goes, are based on a government-like process of using forms for all processes, no matter how trivial. An unreasonable amount of paperwork is a trademark. An overreliance on the process and not people make working for a construction contracting firm a "check the box" exercise. When the paper trail is not followed, questions about the employee's competence and diligence emerges.

A multilayered organisational chart characterises bureaucracy. Several people might have two bosses. Bureaucratic firms tend to be the largest ones in our industry. Quarterly or yearly goals are emphasised. Improved practices and mentoring are less stressed; although the communication of these two ideas is strong, evidence that they are performed is scant. Instead of concern for the firm, career building and compensation growth tend to be sought after by many persons.

From our observations, these firms tend to be built on loyalty rather than merit. Those in power tend to pursue a more extraordinary career (not company) prosperity, so subordinates' and peers' allegiance is sought. One characteristic is the creation of a positive narrative about a person's performance. Loyal insiders retell that narrative.

Since bureaucracies thrive on complicated interrelatedness, clarity is not evident as to who is doing a great job and what practices are working. As a result, contracting companies who suffer this, tend not to be the first choice among sophisticated buyers and job seekers.

The Organisational Culture Continuum

A framework for understanding the development of organisational safety culture - ScienceDirect. Parker et al. (2006) identified a range of company safety cultures. We apply Hudson's spectrum in Figure 6.1 to different ownership cultures of our creation. Culture can be described as 1) pathological, 2) reactive, 3) calculative, 4) proactive and 5) generative. These are sequential. They appear to follow a maturation process in a construction firm's life cycle from our observation. However, some firms never quite reach the generative stage.

A company with a **pathological** culture is one where different company members care more about not getting caught being non-compliant with processes than they do about the results achieved by the company. Your company might have a pathological culture if blaming is common. Examples include the project managers blaming the site staff for not passing in paperwork on time or accurately. The site staff blames the project manager for not ensuring that the requested crews are on-site or that the material is delivered on time. In a pathological company, management believes that mistakes are caused by the carelessness or stupidity of their employees – not by a lax culture. There is a tendency to punish the messenger – that is, employees who speak up about poor processes or habits are censured or punished. When there is a project

The Logic of Safety

Figure 6.1 The "Physics" or Logic of Safety. Adapted from the Heinrich Triangle.

failure in pathological companies, it is not discussed openly but covered up and ignored. It is difficult to action new ideas in this climate, as trust is low.

A **reactive** culture is one where processes are followed in the wake of a poor outcome and where compliance gets more and more relaxed as more time passes when good results are achieved. For example, reactive companies may begin to relax censure or blaming after an incident but revert to old ways when the pressure is off.

A **calculative** safety culture is one where the application of core principles is bureaucratic and rule-oriented. Companies with calculative cultures may have seriously considered the costs of failure and sent their management on a weekend retreat the new (sizzling) idea, which is then passed down from the top management to the rest of the office and the site. A calculative culture may be responding to external pressure from a frustrated owner, but as Hudson points out, the principles are not internalised by employees. Rather, they are applied mechanically as rules that must be followed. Messengers are not shot but might be merely tolerated. Responsibility is compartmentalised such that some aspects of safety are considered to be the responsibility of some, but not others. For instance, there may be rigid ideas about whose responsibility it is to make certain material and crews are never late to the worksite. If it is the project managers' job to make sure these items are timely, no office managers or project site manager would consider it their responsibility to look for or point out potential problems in a calculative culture.

A **proactive** culture is well on its way to being generative but still has elements of a calculative culture. For instance, a proactive culture may keep responsibility compartmentalised but seek to develop processes that allow messengers to be heard and recognised for their proactivity. In proactive cultures, trust increases, but bureaucratic procedures may still trump collaborative processes, hurting trust between employees.

A **generative** culture is one where all employees genuinely believe that success, no matter who gets credit, is in everyone's interest, where employees can speak up about problems without risk and where all believe that good behaviours are everyone's responsibility. In a generative culture, when someone sees a problem or poor practice, that person speaks up and others take their voice seriously. There is no tendency to blame the messenger or cover up when there are mistakes. Employees are encouraged to actively shape the company's practices by suggesting new strategies, enforcing compliance at all levels and participating in planning. Processes are developed inclusively, rather than being handed down from the top management. In a generative culture, success is fully internalised to the point that it is assumed to be present in every process and if it is not, someone will quickly point this out. Messengers are encouraged and supported, responsibility is genuinely shared, improvement is constantly and actively sought, problems result in inquiry and new ideas are welcomed and put into action. In a generative culture, all management believes that problems are their responsibility, not carelessness lower down.

In a generative culture, employees self-enforce good practices because they want to, not because they must. Typically, the leadership of a high performing firm has some charisma and deep character. They can be trusted. Each company member likes their immediate supervisor personally and feels strongly about giving an outstanding effort.

We have found that culture is a primary operational context of any construction firm. This is due to the large input of people's actions. What they do decides the fate of a project's outcomes. The critical outputs are safety, quality, cost and schedule.

These are critical as project outcomes determine the profitability and sustainability of that profitability over the years. Each is affected mainly by people's planning, communication, execution and adjustment. In addition, culture impacts other significant areas of a construction contracting company, such as employee turnover, overhead efficiency and problem-solving.

The Intersection of Common Management Models and Organisational Cultures

Each construction company executive must consider their management model and the company culture. Both of these set each employee's mental defaults or unconscious decision-making.

Your culture helps determine some of these decisions in the way you want them as the construction company executive. But, again, it is a default to which company members act.

We have mapped the resulting dynamics of an Organisation's Management Model and its Culture in Table 6.2. As you can see, companies may have several features, but most of their culture is one of the intersections.

Table 6.2 Comparison of Site and Office Duties and Their Culture

Site	Office
Builds projects with the help of the office	Assists site and administrates the construction of projects
Coordinates the workforce on one project	Coordinates the workforce on all projects
Receives, inventories and installs material furnished by office	Purchases and coordinates material delivery, works with vendors on any delivery problems and returned material
Maintains and utilises equipment	Acquires equipment at the most reasonable price and condition
Reports productivity, equipment utilisation, site conditions and other important information	Uses this information to track this project's performance, discuss with the client and bid next project(s)
Mistakes happen and are seen by all since they occur in an open setting – the job site	Mistakes happen and are not seen by all since they occur in a closed setting – the office
Tied to one job. Lack of a project can mean unemployment	Tied to multiple jobs. Lack of one job does not mean the end of employment
Project success is tied to contract agreements with clients (an office process)	Determines final price and agrees to contract with clients
Directly manages installation process and reports physical progress to office	Indirectly manages and assists physical progress. For example, sends pay requests to clients based on physical progress and then collects payment
Directly controls safety, quality and productivity. assisted by office	Directly controls budgets and flow of information affecting safety, quality and productivity
Critical to have successful site experience	Not critical to have a successful site experience
Measured by metrics established and data collected by the office	Establishes metrics and collects data that measure project performance

As we have spoken to many construction executives, we have found that they desire a team-generative type of company as a group. However, our observation is that it is a rare company that has achieved this.

As a beginning point, it is vital to assess where a firm is currently. This review gives clarity to all company members who will be part of any change needed. Company characteristics and activities are identified and then plans are made to facilitate improvement. Leadership and management engagement are needed to move a company from one environment to another; this planning allows the firm to purposely move in their chosen direction. Importantly, to keep this company atmosphere alive, this clarity enables a company to monitor itself.

Indecision Problems

Some management practices that work well for owner-operator models can be pathological for a team or bureaucratic culture and vice versa. Below, we give some scenarios we have seen that point to some issues that can arise when

Owner-Operator versus Family. Employees can get confused when an Owner/Operator takes on a family characteristic. The leader is a mild dictator of all practices, including final decisions. We have seen an uneven focus on the leader in those firms.

Owner-Operator versus Bureaucracy. A company with clear and enforced policies about promotions can generate distrust when the owner's son gets promoted outside of the established order.

Team versus Family. We have seen team management models conflict with family style bonus policies. For example, a company may have a very informal style for deciding how to allocate the bonus pool, with an amount given to an individual based on the "gut feel" of the owners, but in other respects operate as a team culture, stressing the contributions of teams and participation within teams as the basis of professional success. An employee in a low-performing team may lack feedback about whether the owner understands their contributions and may thus be unsatisfied with the bonus outcomes. Most of an employee's feedback is team-based, but financial rewards are informally decided and communicated, as in a family.

In summary, many leaders and managers think they understand their Organisational Culture. In reality, they have often been surprised at what is revealed when we "look under the hood". Culture is not obvious and what the bosses see may not be the whole story.

All substantial construction firms are informally organised with a Management Model and staffed with dozens of competent people. The method they lead significantly affects their mentality towards the firm and creates the organisation's culture. This culture appears to play a large part in employees' commitment to the company. Commitment is another way of stating the intellectual and physical energy each staff member expends on the company's behalf.

In medium and large firms, many employees are not personally known by the people who hired them. In our intense industry, executives do not have time to get to know staff well. However, leaders can build a culture that serves as a motivator and behaviour guide for all employees. Focusing on work practices, social customs and informal behaviours can transform Organisational Culture.

In our advising engagements, a "team" mentality by staff is often desired by leaders in construction. A Team Management Model enables better-than-peer efficiency if coupled with a Generative Operational Mentality. How to create one is the critical question.

For many construction contractors, the beginning years of their construction business were some of the best. Every person in their company was focused, responsive and caring about company outcomes. Many executives feel they were highly efficient. Also, we have heard the term "electrifying" used. For many of these leaders, repeating that environment is their wish. It was a team and generative culture. Reaching that level can make the construction business rewarding in a contractor's first years.

Whatever the Management Model, it is evident that most companies want to move up the spectrum from pathological to generative. However, we have witnessed companies that have both elements. This can occur because they are confused about their Management Model. For instance, in a team environment, owner-operator or family-type behaviours can be pathological. On the other hand, sometimes it is simply a matter of committing to a Management Model. In another section, we will discuss some of the problems we have observed when companies fail to achieve a Management Model. Finally, we will describe how picking a model has moved these companies' cultures toward the generative.

Safety: Arriving at a Generative Culture

Industrial research strongly suggests that the companies with the best safety records – from oil and gas to aviation to construction – have achieved a *generative safety culture*. This section applies the lessons gained from research across high-risk industries to ours. We describe how a generative culture is more than just a list of rules handed down from management; but rather is a dynamic and evolving set of safety-related expectations and practices that everyone believes in and follows. We describe the four cultural stages that precede a generative culture and tell why an intelligent owner needs to keep moving through these. We explain how construction companies can start the journey down the winding road toward a safe culture and how they will know when they have arrived.

Culture

All companies have a culture, i.e., a set of beliefs and practices shared by everyone. A company's culture is reflected in the management style of the owner (open versus closed door), the interaction between the site and office (whether there is a site-office conflict), the degree to which information is shared across staff (trusting in the team versus oneself) and even the kind of coffee that is supplied in the break room (instant or gourmet, or somewhere in-between). In short, every company has a culture that affects the way employees relate to each other, their leadership and their workplace.

Safety is a part of this culture. With all the different personalities and roles in construction, it might seem hard to believe that there is a single culture that everyone buys into. Particularly with safety. Office staff, who often make the rules about safety in a construction company, have vastly different jobs and roles than the site staff, who must execute safety procedures. It might seem like there is an office safety culture and a site safety culture, or the site has a different culture. Culture is complicated and in some cases, these sub-cultures might exist.

However, it is essential to remember that everyone in a company contributes to its culture. When the office and the site have different ideas about safety, this is part of the culture. When different site crews have different safety standards, this is also part of the culture. The essence of a culture is simply how other people contribute, which makes a culture generative or otherwise.

It is important to note that we have witnessed that company culture does not stop at the project site fence or home office property line. It tends to bleed into all company leadership and staff's personal lives. Our past work tells us that an extreme culture of hyper-competitiveness finds its way into the unguarded intimate moments that we all experience. Of course, when the culture of inclusiveness and steadiness is ingrained into the company philosophy, employees' behaviours are affected in their off-duty hours. The result of either scenario affects their relationships to loved ones.

A Safety Culture

As Hudson points out, there is a difference between "the safety culture" of a company and a company that has "a safety culture". Every company has an aspect of its culture focused on safety, whether it is effective and supportive or punitive and blaming. Companies that allow lax compliance to safety processes have a relaxed safety culture. As discussed in the next section, companies where blame is assigned to individuals for systemic safety issues, have a pathological safety culture. On the other hand, companies that seek to engage employees in the development of safety processes, engage them in blame-free and supportive enforcement of those policies and provide structures for safety processes to adapt when they need to, might be said to have "a safety culture". That is, a culture where all employees believe in the safety mission and seek to support others to comply with and enforce that mission.

Companies should constantly move from "the safety culture" they currently have to "a safety culture". As leadership blogger Tom Northup is famous for saying, "All organisations are perfectly designed to get the results they are now getting". A company with "a safety culture" has designed its safety processes deliberately to generate safe outcomes. In the next section, we discuss one path to a generative safety culture.

A Generative Safety Culture

Hudson and other authors identify a spectrum of company safety culture types, the generative culture's pinnacle. Hudson's spectrum is: 1) pathological, 2) reactive, 3) calculative, 4) proactive and 5) generative. These are sequential. They appear to follow a maturation process in a construction firm's life cycle from our observation. However, some firms never quite reach the generative stage.

A company with a **pathological** culture is one where different company members care more about not getting caught being non-compliant with safety processes than they do about the actual safety achieved by the company. Your company might have a pathological safety culture if blaming is common, e.g., the project managers blame the site staff for not taking safety seriously, or the site staff blame the project manager for not arranging proper safety equipment for being on-site. In a pathological company, management believes that accidents are caused by the carelessness or stupidity of their employees – not by a lax culture. There is a tendency to punish the messenger – that is, employees who speak up about safety are censured or punished. When there is a safety failure in pathological companies, it is not discussed openly but covered up and ig-nored. It is difficult to action new ideas in this climate, as trust is low.

A **reactive** safety culture is one where safety processes are followed in the wake of an accident or near hit. Compliance gets more and more relaxed as more time passes without a safety incident. Reactive companies may begin to relax censure or blaming after an incident but revert to old ways when the pressure is off.

A **calculative** safety culture is one where the application of safety principles is bu-reaucratic and rule-oriented. Companies with calculative safety cultures may have seriously considered the costs of safety and sent their management on a weekend retreat to painstakingly develop new safety procedures, which are then passed down from the top management to the rest of the office and the site. A calculative safety culture may be responding to external pressure from OH&S or an owner to increase safety. Still, as Hudson points out, the safety principles are not internalised by employees. Instead, they are applied mechanically as rules that must be followed. Messengers are not shot but might be merely tolerated. Responsibility is compartmentalised such that some aspects of safety are considered to be the responsibility of some, but not others. For instance, there may be rigid ideas about whose duty it is to ensure that safety equipment is present on the worksite or that staff follow safety procedures. The foreperson and site super-visor's job is to ensure these safety procedures are followed. In a calculative safety culture, no labourers or project managers would consider it their responsibility to look for or point out missing safety equipment.

A **proactive** safety culture is well on its way to being generative but still has elements of a calculative culture. For instance, a proactive culture may keep responsibility compartmentalised, but seek to develop processes that allow messengers to be heard and recognised for their proactivity. In a proactive safety culture, trust increases, but bureaucratic procedures may still trump collaborative processes, again hurting trust between employees.

A **generative** safety culture is one where all employees genuinely believe that safety is in everyone's interest, where employees can speak up about safety without risk and all believe that safety is everyone's responsibility. In a generative safety culture, when someone sees an unsafe practice, she speaks up – and their voice is then taken seriously by others. There is no tendency to blame the messenger or cover up when there are mistakes. Employees are encouraged to actively shape the company's safety culture by suggesting new strategies, enforcing safety at all levels and participating in safety planning. Processes are developed inclusively, rather than being handed down from the top management. In a safety culture, safety is fully internalised to the point that it is assumed to be present in every process and if it is not, someone will quickly point this out. In a generative culture, messengers are encouraged and supported, respon-sibility is genuinely shared, improvement is constantly and actively sought, safety breaches result in inquiry and new ideas are welcomed and put into action. In a generative safety culture, top management staff believe that safety problems are their responsibility, not the result of carelessness lower down.

Whether your company views itself as a family or team, employees self-enforce safety policies because they want to, not because they must. Safety enforcement can be both serious and playful, i.e., a "punishment" system where an employee that has violated safety policy has to buy sandwiches for their team or put a rubber chicken on their truck's dashboard As Hudson puts it, the motto of a company with a generative safety culture might be, "Safety is how we do business [a]round here".

Assessing Your Starting Point

One necessary step on the path to a generative safety culture is to assess where the company is at this moment. As Hudson points out, companies sequentially move through the five stages listed above. There must be a set of safety rules before a robust and generative culture is developed in the reactive and calculative stages. All employees must also share the belief that safety is essential for safety's sake, rather than to avoid injuries, OH&S fines or being shut down. This value of safety for its own sake is a sign that the company is moving from a calculative to a proactive safety culture.

Savvy construction professionals, wherever they are, constantly educate their superiors, peers and subordinates about the importance of safety (as well as the actual practices that reduce risk). However, most employers do not assess the beliefs of their employees concerning safety. That is, most management staff do not take the time to understand whether their team believes safety to be, say, a necessary evil or a desirable foregone conclusion. As Hudson points out, assessing your starting point means identifying the employees' beliefs and assessing whether those beliefs are consistent with a safety culture.

One of the great questions to delve into with employees is the belief that safety and a safe workday are controllable. Some may believe "what happens, happens" without regard to the "physics" or logic of safety i.e., cause and effect. Education (case studies, statistics, stories, anecdotes, graphical models, etc.) can be easily found or created (from the extensive research currently available) to dispel this myth. With the journey metaphor in mind, it is never finished as employees are on multiple journeys, some of which are distracting them from their professional ones. As you know, refocusing on safety sometimes requires a "Safety Stand Down" and is necessary even for firms with stellar records.

Figure 6.1 is a simple example of the "physics" or logic of safety. We have never found anyone to argue against the link of cause and effect of behaviours (safe or unsafe) leading to incidents and fatalities. This is most likely understood by your experienced or savvy employees, but possibly not the least experienced. Unfortunately, those with little understanding can hurt others as well as themselves. This is a sad reality. All company personnel might be educated (and tested) above the legal requirements depending on your culture.

Making a Plan

Once you know where your company will start its journey toward a generative safety culture, you can develop a plan to move along the winding road. As Hudson points out, the road is winding because it requires iterative contemplation, preparation, action, and maintenance. However, in construction, having a plan gives employees a focal point to establish the foundation of a safety culture. This plan will have certain critical elements, including transparent safety processes, plans for dealing with input and new ideas, distributing responsibility and other urgent actions. In addition, the plan itself will need to be updated periodically as the company's safety culture moves through the phases.

Any serious safety plan should name the president or chairperson as its leader. The plan needs that kind of authority to make it the first thought of each employee each day. The only drawback is that this high-ranking executive does not demonstrate a singular commitment to safety's importance.

It is important to note that safety in any creditable university program is not an elective but a mandatory subject. Additionally, in some countries such as Australia, executives can go to prison for an accident resulting from a lax or inept safety program. Therefore, a safety plan must have a demonstrative component i.e., an action plan with measurement and adjustment.

A serious plan starts at the project site. It assesses and analyses many areas such as culture, practices and data. It goes above OH&S requirements. It may even purposely infuse a culture of everyone "looking after each other".

Knowing You Have Arrived

How will a construction contractor know when they have arrived at a generative safety culture? The word arriving is a misnomer. A company will never "arrive" at a generative safety culture – maintaining everyone in that culture takes commitment and diligence. By definition, culture is dynamic and shifting. However, at the end of this journey, leaving things well organised with much data, processes and anecdotal capture for the next generation of company leadership appears to be the best goal.

No culture that ever lasted did so without changing. In construction, there are always new clients, new types of equipment, new employees, new management structures, company growth and other dynamic factors that make safety a moving target. A construction company is on a journey by embracing a culture of innovation, iteration and generation. Some of it is uncontrollable. However, for the controllable part, the question that may haunt some in the industry after a negative safety incident, *"Did we do all that we could have done?"*

Construction Craftsperson, Equipment Operator and Labourer Shortage

The current construction workforce shortage is reaching epidemic levels. Although much concern has been expressed, the problem persists. This inadequate workforce is a barrier to creating better infrastructure, adequate housing, upgraded utilities, improved public spaces and job creation via private industry and public investment. Owners, also known as Construction Service Buyers (CSBs) or Construction Users (CUs), understand that contractors are a crucial bottleneck in this country's-built asset process. We offer five areas of focus in this section to prompt discussion of how CSBs can help to solve the workforce shortage in construction.

A Collaborative Solution

We ask CSBs to consider reinvesting in the industry in the spirit of enlightened self-interest. All highly respected firms reinvest for future opportunities. Look at the list of top-performing companies in any industry – what started out as reinvestment evolved into harvesting opportunities while continuing reinvesting. If this workforce issue is left as is, the numbers of quality construction people will continue to fall short of demand even as the industry evolves into a technology-centric one.

We know many CSBs interested in a robust construction industry and we offer the following suggestions for those organisations looking to be part of the solution.

1 Generative Industry Hiring

We suggest a generative system of recruiting, hiring and onboarding. As you may know, a *generative system* builds upon its internal resources. In this case, the funding of educational buildings and infrastructure might be the place to start. Suppose educational contracts require that a certain proportion of workers be matriculating or graduating, for example. In that case, educational organisations become enablers of a future capability that serves the industry and themselves.

2 Formal Onboarding

We also recommend that CSBs have requirements about the onboarding – first six months) of new site employees – including but not limited to those who come through the generative industry hiring processes. It is clear to us that a company employing this new worker must have an onboarding program planned and ready to execute. Most knowledge in construction is learned on the job and critical gaps in knowledge can produce significant inefficiencies and bad habits. Reasonable standard operating procedures (SOPs) can smooth the way for new hires and longstanding staff alike. Again, this requirement to hire might start with educational CSBs.

3 Steady Flow of Work

We also recommend that CSBs partner with construction firms rather than establish the adversarial relationships currently expected in the industry. To us, the "too hot, too cold" nature of construction leads to two problems: 1) rushed hiring and training (too hot) leading to incomplete learning causing increased safety risk and inefficiency; 2) layoffs (too cold), resulting in lost technical knowledge and enthusiasm for the industry. In our opinion, both are symptoms of an unsteady design and procurement process. This may result from construction service buyers, funders and end-users not realising that a steady design and procurement stream allows contractors to plan and commit resources. As a result, layoffs and chaos drain enthusiasm and discourage self-improvement by the worker. Conversely, a predictable and uninterrupted project flow allows workers to build on previous learning and buy-in emotionally to their company and the industry.

4 Formalised Planning

Standardised project planning would also help existing labour be more productive. A technical and humanistic pre-mobilisation process would provide consistent benefits to justify its expense. We know from much research that early planning has a high payoff in adhering to budgets, project timelines and reducing interpersonal conflicts.

As an example, a CSB may set aside 5% of the overall schedule and budget to compel all project stakeholders to produce a plan that includes:

1   a technical understanding of the project.
2   a path of project execution with contingency planning.
3   a demonstration of team cohesiveness.

Achieving the above could be led by the prime contractor. Since it requires creating deliverables, the cost and time should be spelt out in the contract. Thus, the expense and days needed for this effort, including all-inclusive stakeholder participation – Site personnel especially are critical attendees – would be appropriately compensated and placed in the normal schedule, providing an incentive to do it very well. The deliverables could be a few or many such as a perfected Virtual Construction Model, project planning document, CPM schedule and project letter of instruction. Done well, each attests to a full review of all possible construction approaches and appointment of the best way for the risk involved. Factors the deliverables would address might include site conditions, plan and specification conflicts, material availability and staged use requirements.

5 Facilitate Data Collection

Our long view is that focusing on data will help the construction industry evolve from one based on opinion to one based on fact. Workforce practices data could help refine good practices. Also, those that are ineffective could be identified. Between the enhancement and deletion, it would be a nice windfall. This type of information will make the business more manageable and predictable. There is a dearth of data in construction, so the industry is forced to rely on opinion to make crucial decisions. As you may know, some beliefs are based on incomplete experience and rarely do the past problems perfectly fit the current situation.

In conclusion, our industry needs new workers who are well-trained and enthusiastic. Unfortunately, the current adversarial relationship between CSBs and contractors prevents us from using collaborative strategies. However, a change in approach will encourage focus between contractors and construction service buyers to increase new and potentially career-minded entrants. Importantly, it is not enough to bring people into the industry. Once they are employed, significant onboarding must be planned and executed to keep this targeted group engaged and improving.

Reframing the worker shortage problem, including CSBs and contractors as collaborators, should help. Both parties will win long-term with a well-developed and larger work workforce.

Learning and Forgetting in Construction

Learning in Construction is a correlation between proficiency and practice repetitions. If the same task is executed often in time, ability (learning) improves. Conversely, when interrupted or spaced longer in time, proficiency degrades (forgetting). Intuitively, construction firms know this is the root cause of safety, quality and productivity problems.

It is evident that the learning organisation has a high probability of continued success. Many business leaders have commented numerous times. "The pace at which people and organisations learn may become the only sustainable advantage" Ray Stata, CEO Analog Devices Bill O'Brien, Chair of the Hanover Company, states, "In the learning organisation, the new dogma will be vision, values and mental models" (Senge 1990).

Construction learning is the focus of construction firms for staff. Butcher (2006) careful construction of diagrams help build mental models with which the employees may be used to solve problems quickly or interpret a new situation accurately.

Simple diagrams are more efficacious than written instruction. More detailed graphics appear to confuse or overwhelm.

Straight forward graphics appear to generate inferences about the content. Prompting more inferences is a feature of deep learning. It seems that instructors planning for time for a course should be partially dedicated to creating accurate diagrams. In fact, multiple graphics that illustrate the same concept boost understanding. The efficacy of visual representation centres around the comprehension processes that natural to learning. Attending to the cognitive power of graphical illustrations is another step toward refinement of teaching construction management.

How might a construction firm think about managing the reality of learning and forgetting? Research shows that whenever a task is repeated many times in construction, there is usually a corresponding reduction in how long it takes to perform. This is referred to as the learning (or experience) effect. The task is repeated with minor interruptions while a simple mathematical function can approximate the curve. It is a negative logarithmic curve whenever there are significant interruptions: some of the learning is lost and performance regresses part of the way back up the curve; this is referred to as the forgetting effect.

- There are many implications for the learning and forgetting effects.
- You can plan for work to accelerate if it is repetitive.
- Thus, fewer crews may be required to get the work completed on time.
- You should not swap crew members and you should avoid changing crews since the benefits of learning will be lost.
- The learning curve will remain flat in the worst-case.
- If work is falling behind schedule, you should add crews (or crew members) early so that they have more time to learn.
- This way, fewer crews will be needed and will operate at higher productivity (so that costs come down).

The learning curve can be approximated by a negative logarithmic function.

- For every doubling in the number of repetitions, there is a constant decrement in CAT (the Cumulative Average Time to complete those tasks)
- $CAT_n (1/n).(t)$

An enabling learning environment may be another form of management intensity. The lack of disruption, including a quiet environment for a person to focus, allows for comprehension.

Common Mistakes

In the rush of completing projects, all stakeholders may hurt learning. Many do so unknowingly.

The project they manage is only focused on another person's responsibilities for some construction professionals. Unfortunately, this sole priority can make managers interrupt others.

Critically, most construction firms do not consistently hold post job review meetings to clarify and document lessons learnt. Our industry experience informs us that revisiting a project's events, struggles and successes is not desirable to most people, especially when a new project is waiting to be started. This is lost learning. We believe hindsight is invaluable even though painful at times. Furthermore. To hold a meeting with all department managers attending affords learning and teaching opportunities. It is important to remember that construction contracting is the business and all departments should serve the primary function.

Post job review is essential for a "learning" organisation. Consistent project review sessions that are organised, unemotional and objective will be true learning events. Using a third-person attitude – "we didn't" (versus Jim did not) keeps you from the blame game. Positive facts, as well as negative ones, keep your review balanced and energised. From this, the beneficiaries are many: the company, the individual and the project teams (Figure 6.2).

The Learning Curve is Unique to Each Person, Crew and Company.

As companies seek to improve outcomes, the human factor must be part of the process. We see it as the largest single input to successful construction contractors. Therefore, understanding more about the people who generate outcomes cannot be ignored.

Indeed, people become tired and forget even though fully immersed in an activity. Safety accidents are more serious also as people work past 40 hours a week.

In conclusion, the learning-forgetting dynamic significantly affects the construction industry. Its stakeholders benefit from the proper use of it. However, it is one of many aspects of constructing a project.

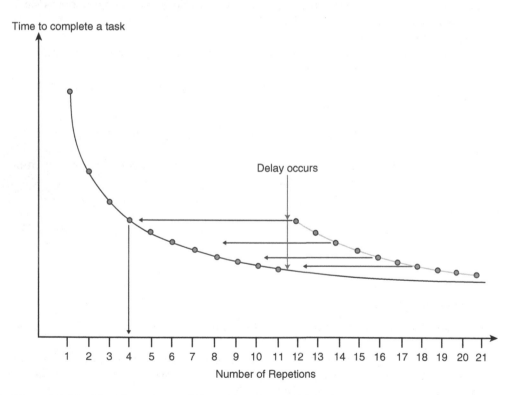

Figure 6.2 Model of Learning and Forgetting Curve with Delay.

Source: University of Florida BCN 5618 Scheduling (2010).

The Forgetting effect represents a risk factor to be managed. Each person on the project, whether construction services buyer, design profession, construction employee, or supplier staff, should be well-versed in this issue.

Learning and Forgetting appear to be one root cause of safety, quality and productivity issues. This is one reason we recommend contractors not to expand their business aggressively. Too many new people to teach (18 months for them to assimilate all your practices and expectation) and too many people to lay off (cost of training is lost with little return).

A Job Description Approach

The best job descriptions in most construction firms are those written to facilitate job performance measurement. The measurable areas are not general such as "is team player", but specific terms such as "plans in writing each day of each job they build" This communicates and drives above-average behaviour.

Companies that give authority to employees at the lowest possible level tend to be less bureaucratic and, thus, faster in building work and resolving problems. They also never publicly second guess, although they may do so behind closed doors. To be clear, authority and responsibility should be given in equal amounts while supporting the actions of those involved.

Project Manager (1 of 2)

Reports to: President

Supervises: Subcontractors, Suppliers, Service Providers, Employees as designated

Works directly with Company's Clients

Basic Function: Plan, organise, direct and control the building/renovation aspects of the company and jobs in such a way as to meet company objectives of costs, quality, schedule and safety.

Duties and Responsibilities:

1 Hire and Train:

    a Hires needed personnel.
    b Oversees the coordination of personnel.
    c Establishes and implements a new hire orientation process.

2 Participates in establishing company objectives:

    a Participates in establishing goals and objectives for the construction area.

    b Gain buy-in on company objectives from subordinates.

3 Planning:

    a Participates in the pre-job planning process.
    b Participates in establishing goals and objectives for the custom building/ renovation of the company.

---

    c   Oversees planning of Building/Renovation and use of company re-
       sources.

    d   Assures the planning process is followed.

    e   Prepares job descriptions for subordinates.

    f   Coordinates and plans with Clients.

4 Organising:

    a   Oversees the organisation of the field.

    b   Sees that administration of the jobs is effective and efficient.

    c   Sees that the field is well organised.

    d   Makes recommendations for organisational changes.

    e   Helping with Client demand such as access or storage space.

5 Directing:

    a   Troubleshoots complex construction problems.

    b   Establishes ways to motivate and keep personnel/subcontractors/sup-
       pliers and service providers' morale at a high level.

    c   Reviews subordinates periodically and makes recommendations for
       improvements and wage adjustments.

    d   Gives directions to subordinate managers.

6 Controlling:

    a   Sees that construction cost targets are met.

    b   Helps establishes strategies and action plans to beat the budget.

    c   Establishes action plans and oversees such strategies as to achieve our
       goals.

    d   Makes sure job cost feedback is furnished to subordinates.

    e   Oversees the terms and conditions of construction agreements.

    f   Oversees job schedules and makes sure that they are updated and
       reported.

    g   Keeps Client's satisfied, but not at the substantial detriment to the
       company.

Key results:

1 Return on Investment.
2 Customer Satisfaction.
3 Subcontractor Coordination.
4 Professional Appearance.
5 Adherence to Deadlines.
6 Growth of subordinates.
7 Independence.
8 No Rework of PM Tasks.
9 Teamwork.
10 Improvements to operations.

The president will grade the project manager's performance on each factor of the above. He will use a 0%–100% scale. The resulting cumulative average will be multiplied by the total maximum incentive pay per completed home. The resulting will be the bonus earned by the project manager. It will be paid 90 days after completion of punch list and closing of the project, whichever is later. If the project manager leaves the company's employ before this date, then incentive pay is not due.

Performance Review: Project Manager (2 of 2)

Name: _____

Date: _____

QUALITATIVE

		1=Fail	*2=Poor*	*3=Satisfactory*	*4=Excellent*	*5=Stellar*
1.	Return on Investment	1	2	3	4	5
2.	Customer Satisfaction	1	2	3	4	5
3.	Subcontractor Coordination	1	2	3	4	5
4.	Professional Appearance	1	2	3	4	5
5.	Adherence to deadlines.	1	2	3	4	5
6.	Growth of Subordinates	1	2	3	4	5
7.	Independence	1	2	3	4	5
8.	No rework of PM Tasks	1	2	3	4	5
9.	Teamwork	1	2	3	4	5
10.	Improvement	1	2	3	4	5

GOALS:

_____

_____

_____

RECOMMENDATIONS FOR IMPROVEMENT:

_____

_____

_____

Meeting Management: A Needed People Skill

"meetings will continue until productivity increases". We have all lived this saying. Meetings can be timewasters. However, done correctly and with discipline, meetings are the only way to gather and disperse information between two or more persons. Additionally, they force people to give non-verbal hints as to what they think. In-person, more is communicated than in an email.

Most construction people dislike meetings. This is because they would instead be building something than talking about it. Hence, being efficient in time, such as time discipline and content focus, will make you more popular with others.

Meetings are essential in the responsibilities of a project manager. Meetings are often the vehicles used to communicate information and make crucial decisions. Obviously, productive meetings are critical to a project's success. On the other hand, poorly run meetings reflect on you and the company concerning how organised the project seems to be and the degree of importance that is placed on it. As the primary facilitator, you are responsible for providing leadership and direction in the meeting. Below are some guidelines in running effective meetings:

Guidelines:

1   Every meeting must have a plan. An agenda will guide attendees on the purpose and objectives for the meeting.
2   Communicate the agenda at least 24 hours before the meeting. This allows people adequate time to prepare.
3   Invite only those who need to be there and can make decisions. Any others will feel the need to talk to justify why they are there and waste valuable time.
4   Use the agenda as your guide in running the meeting efficiently. Have someone else take the meeting notes. This allows you to concentrate on managing the meeting dynamics.
5   If there are assignments made during the meeting for follow-up, ensure that each person knows what they are responsible for and when to be completed.
6   Determine which items are to be discussed at future meetings. This may include items in which you do not have adequate information in which to decide.

We worked with a construction firm and asked them about when things not to do in meetings. In our discussion, we list 20 items that they labelled "Cardinal Sins". We list them here as specific actions not to do and if you are the chair to discourage:

1   A meeting did not have a sign-in sheet.
2   Reusing an old agenda.
3   Intentionally raising a substantial topic that is not on the agenda.
4   Too many meetings; meetings interfere with operations.
5   Late start or finish.
6   Losing focus or going off-topic.
7   Not engaging audience.
8   Not being prepared – doing homework the meeting.
9   Using devices to communicate while in meeting
10   Not communicating information to people who missed the meeting.
11   Discussing someone who is not in the meeting.
12   Not defending junior employees in meeting.
13   Not insisting on a respectful tone.
14   Allowing escalation of emotions.
15   Not allowing everyone to contribute.
16   Feedback not provided when requested Unnecessary attendees.
17   Excluding key contributors.
18   Inefficient meeting times – Monday morning or Friday evening.
19   Not following Robert's Rules of Order.

There are many types of meetings. Each has to be managed differently.

Project Turnover Meeting – Estimating may run this meeting. This is one of the critical meetings in the entire life of a project. All project team members and estimating participants including client development involved, should be present.

Project Coordination Meetings – weekly meeting with project partners to encourage them to work together and develop "perfect information" with which to construct. They should discuss the priority of construction space and possible conflicts that may exist. Some companies may employ a Mechanical Electrical Plumbing (MEP) coordinator to assist with this process on larger or complex jobs.

Project Review Meetings – The project will run weekly or bi-weekly project review meetings. The purpose is to update management on the job's progress and draw upon the experience of other managers in the company for assistance. The project manager prepares the Project Review Meeting Package for the meeting. It should include the following items:
"Hot" issues list.
Budget review.
Buy-out status.
Project schedule.
Subcontractor status report (if appropriate).
Change order log.
Request for Information (RFI) log.
Shop drawing log.
Delivery schedule and procurement issues.
Weekly Subcontractor Meeting – The project manager facilitates meetings with representatives of major subcontractors who can commit resources (manpower and money) for their companies. The project manager should generate minutes and distribute them within a reasonable time limit (two days) of the meeting. The meeting covers the following topics:

Human resources.
Delivery of materials and equipment.
Change order pricing and requisitions.
Project schedule.
Submittal status.
Coordination between trades.
Action items to be accomplished by next meeting.

Set the right example by starting these meetings on time and use your agenda to control the meeting. This will enable you to move the meeting along at an appropriate pace. One of the biggest complaints by subcontractors is that these meetings "take too long" and "waste a lot of my time".

Client Meeting – This meeting aims to review the project in general, discuss progress and what actions are needed to improve progress on the project. It should also be a forum for bringing up issues, identifying problems, receiving input from the architect and coordinating various trades' interactions. If the architect is responsible for preparing the minutes from these meetings, read them carefully and review for any discrepancies you may find. Then, respond immediately to those discrepancies in writing. Here are some of the agenda items to discuss.

- Job progress – Overall and two-week look ahead.
- Variations.
- Outstanding requests for information.
- Overdue requests for approvals.
- Interaction between parties and requisitions to be approved.
- Action items noting parties responsible.

The last item, Action Plans, are part of every meeting. We are sure you had experienced this frustration – after sessions when you had a strong agreement on the course of action only to have people forget their commitments to you. That commitment was counted on for you to meet your deadline. You are expected to deliver information to others based on the meeting.

Several phone calls later, your time is wasted reminding people to complete their tasks. It is not productive, nor is it profitable. Moreover, it can be embarrassing when you lack promised information to keep your promises.

We have found an effective technique to keep others (and yourself) focused on the actions that need to be completed.

There is no more basic management practice than the use of an action plan template. Many contractors and their field supervisors and project managers have found it is an effective way to document what has been agreed and thus, keep people committed to necessary actions.

A repeated offender can be faced with the physical evidence of unachieved commitments. It is a fact placed in writing, leaving personality and emotion out of the discussion.

Any print shop will print and bind these in a gummed, carbonless pad. It is productive for carrying in your briefcase and using it as you meet with people daily. As a meeting or discussion occurs, agreements to perform specific tasks will appear. Document them, give a copy of the action plan to others and keep one also. It is simple and effective (meaning fast and efficient).

We should expect others to keep their commitments. We know we must keep ours. It is one of the reasons that people buy our construction services. However, with others, sometimes we have to force them to keep theirs.

A People Parable: The Rose Garden

Here is a parable that may help partially frame the Human Resources issues in a construction contracting organisation. It is meant to clarify people selection and promotion barriers and opportunities.

A gardener has twelve roses in his garden. Unfortunately, like all managers, this person has limited daylight and resources such as water, fertiliser and the time that can put forth to grow these beautiful plants.

The roses are not all the same; some have a better genetic makeup and can, if nurtured, become grand champion roses. Others are good, not great and some are just ordinary.

One approach to this garden exercise is to divide the time and resources by 1/12th to allow fair distribution or equality. It seems to be the best way to handle the situation. Or is it?

Given that roses are a competitive business and that a single rose plant may be a leverage point for national recognition, the gardener should carefully consider their approach. National recognition can bring more resources to the gardener and benefit all the roses and him.

Meaning: As with roses, no two people are the same. Some pay close attention to you and your expectations. They are better than average. In the future, those same people may be a leverage point for others. In our business, great labourers can become great journeymen and operators who can become great managers who can become great executives. Our people need to know that spending time with talented and motivated folks is what we want.

In construction, to ignore focused and talented people and give them an equal amount of our time and energy is like the rose garden problem. If our managers do not understand growing the best of our people, we will always have above average but not great leaders. These above-average leaders will produce and deliver a good budget, schedule, safety and quality results, not great. Great is always the goal.

Some people will reach the limits of their growth. That is nature at its clearest. Not everyone wins the highest honour. That should be no shame to anyone. The story does make people think about this reality.

7 Retiring From Your Business

As Stephen Covey (1999) famously said, "Start with the end in mind". Thoughtful strategic planning might make that the beginning point. We have observed that the blind spot of many improvement efforts is the lack of connection to a firm's owner(s) retirement. Some transformational programs' "flavour of the month" nature is poison to serious and disciplined improvement.

It is critical to note that the value of a construction firm as an asset ascends or declines depending on the predictability and profitability of its performance. Both of these factors are considered when assessing the market price for ownership. As you will read, many times, the last three years of performance are deemed to be indicative. So from the previous chapters, building a smart culture and hardwiring practices into the computer system can achieve high levels of consistent performance.

The Last Management Act: Selling Your Contracting Firm

Owners of contracting firms are all eventually faced with the inevitable. They will not be leaving this earth alive and breathing. Only astronauts do that consistently, but then, they return. At some point in their careers, contractors' energy and attitude will wane. They will see that the business they built needs new leadership. Their body and mind tell them that the time has come to pursue a different role in life.

When that time comes, what do they do about transferring ownership of their firms? That is often a difficult decision, both intellectually and emotionally. Fortunately, several options are open to them.

Sometimes, it is just a matter of legal paperwork giving ownership to a son or daughter with a family payout agreement. This is a common occurrence in the construction business.

On occasion, employees will step in as the new stockholders. But, again, an agreement must be sworn to and a cash payout schedule accepted by both parties. In such transactions, if a talented core group owns most of the stock, it can give the company a new life and a thriving existence for many years.

The most complicated and daunting option is the third party buy/sell agreement. In many cases, the buyer and seller will not personally know each other beforehand. This leads to a longer transaction process with an unpredictable ending.

In all these examples, the owner's goal must be to accomplish two things:

*   Change their relationship to the firm.
*   Derive some monetary value for the firm.

DOI: 10.1201/9781003290643-7

The transaction between the former and new owner(s) will be unique. Each of the parties in any buying/selling situation will have unique sets of circumstances and needs.

As stated before, emotions are part of the equation. The owner who has spent many years nurturing the firm now has to say goodbye. As you know, this is not an easy transition. A significant number of proposed transactions do not go through because of this very reason.

The selling party needs to think about this life change and what they will do in the future to address this issue. Having a specific plan for particular activities to fill the time is more beneficial to make the transition than just thinking about "enjoying retirement".

Once the seller achieves a comfort level with the idea of divestment, then the next hurdle is to think through the transaction from both the seller's and acquiring party's perspectives. All contractors have a relatively clear idea of what they want to happen. However, the buyer's perspective is just as crucial for the seller to keep in mind. Ask yourself, *"What would I want or accept if I were buying a contracting firm?"* Maintaining a balanced perspective in your approach to the negotiation will not kill it prematurely.

Do understand that construction firms do not sell at a premium. While in some rare cases, they may be valued at as much as four times earnings (plus the value of the firm's assets), most firms are typically valued at no more than twice earnings. Sadly, this is not a great amount of money for the years of effort invested in building a profitable construction organisation. As you can guess, if financial concerns are your top priority, then keeping your construction firm until you die makes the most sense. But if your goal is not strictly a financial one, then selling your firm may be the best move, as some of the best benefits cannot be valued in dollars.

It is essential not to overlook that you must have a business to sell one. A valuable company executes well-organised processes by employees who work independently on most tasks. Simply put, a business can operate for short periods without the owner present. A job disguised as a business is one where the owner must be present at all times for the company to operate. A profitable business has value and can be sold at multiple earnings plus asset value, whereas a job is worth not much more than the assets.

Contractors buy contracting firms. A majority of transactions occur between professionals. Yes, sometimes there is the one-off transaction far afield. For example, money management firms and public utilities have purchased construction companies. However, these are rare cases. In essence, only contractors want to be in this business. Some say nine out of ten. Hence, the search for a buyer is more efficient if pursued within the population of contractors.

The Process

1 Gather Data:

  1 Tax returns.
  2 Financial Statements.
  3 Closed Job Information.
  4 Organisational Chart.

    5   Loan Agreements.

    6   Employment Contracts.

    7   Other important documents such any legal settlements and supplier agreements.

2 Conduct Interviews:

    8   The business broker will interview the owner(s) after studying the corporate information. This needs to be accomplished so the broker knows all circumstances surrounding the company and can answer them for a prospective buyer.

3 Assemble Contact list:

    9   The business broker will assemble a potential buyer list. The owner(s) will review the list, remove those objectionable people and add people they may know.

4 Make Contacts:

    10 Typically, formal letters will be sent to individuals who have been identified as potential candidates. The notes have a sales flavour and describe the company and its situation in general terms, so there can be little chance of someone guessing which firm it might be. After the letters are sent, follow-up phone calls are made to generate interest.

5 Confidentiality and Non-Disclosure Agreements:

    11 Once one or more potential buyers signal interest, formal documents memorialise the interested party's pledge not to disclose or otherwise disseminate your company information, including the fact that it is for sale. A good agreement will have remedies for violations. The treatments can be financial penalties.

6 Send Company "Book" of Information to Potential Buyer:

    12 The business broker will create the book of information. It will disclose the details of the business, its financial status, market share, organisation structure, labour situation, among dozens of other areas. The buyer should be able to decide whether the company is of interest based on the data provided. If not, the book should be returned.

7 Face-to-Face Meeting between the Seller and Potential Buyer:

    13 This meeting is crucial to establish a working relationship and to see whether the chemistry is right between the parties. Unfortunately, sometimes this is the last meeting between the two and the negotiation dies. However, if both parties feel there is a value in moving further, additional session (s) will be held.

8 Potential Buyer Issues Letter of Interest:

    14 Once a letter of interest is sent to the seller, some steps occur in rapid succession.

9 Due Diligence:

    15 The buyer will have 60 to 90 days to inspect the business being purchased. They will complete activities such as reviewing the books, inspecting the fixed assets and auditing the projects.

10 Create Purchase Agreement:

> 16 The parties will create a purchase agreement that spells out what is being bought and under what conditions it is being accepted at what dollar amount.

11 Set a Closing Date/Hold the Closing:

> 17 The business broker, the seller and the buyer will set a closing date that is reasonably close in time – 90 to 120 days – from the date of the Letter of Interest.

Types of Buyers

There are three types of buyers. You should be aware of each type's goals and focus:

- Strategic Buyer.
- Economic Buyer.
- Familial Buyer.

A strategic buyer is someone looking for geographic or service expansion. They can be looking in your state. People from the north sometimes buy a firm in the south. Alternatively, a buyer could be looking for a business to expand their service offering. For example, a plumbing firm may purchase an HVAC contractor. Suffice it to say there are numerous possibilities.

An economic buyer is looking for a reasonable price for the value received. The potential buyer might speculate that they can buy the firm at X and hope it will be worth 2X in a couple of years. They will sometimes purchase a firm in financial trouble to capture its workforce. As you know, journeymen are not easy to grow.

A familial buyer knows the owner through a family relationship or is the company's employee(s). (Company employees are often considered an informal family). This kind of transaction is a good first option. People are always more comfortable on both sides with this situation. People know each other well and the payment terms often are friendlier. Trust is typically high.

The Value of Your Firm

The value of a construction firm is relatively simple:

> A multiple of the predictable income stream plus the value of the assets ascertaining and agreeing on a predictable income stream can be complicated, historical performance notwithstanding. A commercial or industrial contractor is based on several factors, including bids and present and future work in process. A residential builder might be found on inventory, name recognition, product and economic factors. A service contracting firm is straightforward since the business acts more predictably.

A common mistake is to believe that an abnormally great year can be used as the basis for value. It should not be. An average over several fiscal years should be used.

The market determines the multiple and the value of the assets. The last few years of closed transactions are a reasonable basis for valuations. Certified professionals in this area are the only ones who can ascertain a fair market price. In some cases, each party will hire its business appraiser and then negotiate with the other with its report in hand.

The owner has a range of working options after the sale. Some, who know more than anyone else about the firm, stay involved as the leader for a period. This gives a smoother transition to the new owner, who will need to become familiar with the business.

The other end of the spectrum of options is for the exiting owner to leave the business altogether as soon as the sale is complete. As you might surmise, there must be some transition period, but the transition time is brief in this scenario.

The seller should expect to be compensated for time spent working with the new leader(s). If the previous owner stays on as a senior manager, then an "earn-out" is formulated. Typically, it contains a modest salary plus bonus. The bonus is a percentage of profits, usually collected gross margin.

Some Traps

- It is estimated that most of all proposed transactions do not go through. There are several reasons for this, and they are not surprising.

  - An owner feels the company is of higher dollar value than the market will bear.
  - The candidate company has no second-in-command to take over. The new owner most often has no one to come in and take charge. They are looking to the company for leadership. If there is none, the risk is too significant to the purchaser.
  - There is a "profit fade" between the initial contact campaign and the closing date. After that, some owners start leaving the business mentally, and the financial statement shows it.
  - Accounts receivable can be a contentious issue. The uncertain nature of construction billing, i.e., bad debt, variation disputes, over/under billing and unresolved claims, lead to a discount of the amount showing on the A/R register. Some buy/sell agreements have separated classes of receivables from each other and calculated different percentages for valuation purposes. In one case, a local contractor negotiated a deep discount then, after the transaction closed, proceeded to collect over 95% of what was due. As you can guess, he was a well-respected and forceful person.
  - Many times, real estate is of little interest to the acquiring party. In negotiations, this may be true; other times it is a ploy. Be aware that unless you have a unique office site and yard, lower total cost real estate is available in your area.
  - Delay in the closing date. Impactful events occur more often. Therefore, an unnecessary delay raises the risk of one of these events sabotaging the transaction.

- An estimated 80% of the transactions that do go through are sales to employees or family members. Neither can make a cash payment for the business, so a payment plan must be formulated. This means the acquirer will be using your future profits to buy your business. If they fail, you may get your business back.

One caveat to anyone considering selling their firm: There is no such thing as "retirement". Understand that you must be intellectually engaged in something. To not do so decreases your chance at a long post-contracting life. Additionally, if you talk to others who have sold their companies, you will find happiness in doing something challenging or rewarding.

Make no mistake. A construction contracting firm is difficult to sell. It takes several factors to come together simultaneously. Most importantly, the two sides must be willing to negotiate and be reasonable in their demands. Other factors that complicate the matter are valuation, economic forecasts, business performance and the regulatory environment. But, bottom line, if the two parties want to, they can make the transaction happen.

An End of the Journey Parable: Two Feet of Water

A swimmer started swimming across a great lake. Their goal was to finish to the other side in record time. She wanted to do better than anyone else.

She started early one morning at daybreak for the journey. Her coach was at the shore waiting for him when he began. The swimmer had trained hard and was in excellent shape and she was focused on the goal he had set for himself.

She started the long swim; she was fast and straight. She covered a sizable distance by noon. It would be several hours before finishing, but her confidence was high. The swimmer continued ahead.

Currents, underwater obstacles and even a small mechanical breakdown by her chase boat delayed her. He was starting to get frustrated. These things she did not plan for or expect. They should not have happened but were no less real.

It was getting dark when he was within reach of the shore, only a mile away. However, the swimmer started to cramp. Their muscles were tired and any stored energy was long gone. It was starting to become a matter of will. Their coach could see their swimming form change slightly and knew she was getting to the end of their capacity.

At a depth of two feet of water, the swimmer could not go on. She was spent and could not even stand up. The swimmer was delirious, exhausting all the nutrients in her body, affecting everything, including their thinking.

If he were alone, she would have drowned in two feet of water. Thank goodness the coach was there. He pulled him out of the water. There would be no record swim today, but he survived.

A week later, the coach met with the swimmer to talk about the experience. The coach was much older and knew that there were lessons to be learned from this.

However, she only asked questions, such as:

- Is reaching for a world record wise when trying something for the first time?
- Do boats break? Are there obstacles that no one can know of until we come across them?
- Is considering more experienced people's opinions important when attempting something challenging?

Did You Realise You Could have Drowned in Two Feet of Water?

The swimmer was silent a first but did answer. As she thought then spoke and discovered the meaning of the questions.

Meaning: We cannot exist alone. We need others to interject and help us at times. Without doing so, we would make great mistakes in our judgment (attitudes and thinking) and in our actions (doing).

Questions prompt thinking. Statements close thinking off. Even if you know the answer, let others discover it. Once they do, it is theirs to keep.

As shared in the parable, the coach rescued the swimmer and pulled him to shore. Without the coach, the swimmer would have perished. She certainly would have been underwater for several minutes. She would have suffered irrevocable damage and it would have affected her the rest of her life. Thus, having someone there (an extra set of eyes) in any critical endeavour cannot be ignored.

8   Legendary Management Book Reviews

This chapter illustrates one need for the construction industry: applying popular and general business books to construction contracting's dynamics. Each of these books has valuable insights, but each contains fewer practical recommendations and is sometimes dangerous to the health of a construction firm.

The books reviewed have reached some level of notoriety in the industry, and we felt that there is value in reading, understanding and analysing them.

Great by Choice: Uncertainty, Chaos and Luck

Collins and Hansen's book, *Great by Choice: Uncertainty, Chaos and Luck – Why some firms thrive despite them all* speaks to the construction industry's unique set of conditions. Overall, our industry's risk/reward dynamic is like no other. We have a fast-changing and risky environment that is complicated by its custom nature with many uncontrollable forces. The focus of this section is to analyse *Great by Choice's* conclusions and analyse them as possible answers to construction contracting's challenges.

Our firm has assessed three of Collins et al.'s most notable of books: 1) *Good to Great* (2010) 2) *Built to Last* (2016) and now 3) *Great by Choice*. These works are focused on business performance and sustainability. We believe that *Great by Choice* is the most applicable of the three books to construction contracting. We see many operating insights and a focused number of transformative concepts as extensions of our most efficient clients.

The book's research framework set a high-performance threshold – firms that achieve ten times (10x) better performance than an able competitor in a changing and risk-filled market over a 15-year period. The research team found the following industries had 10x performers in extreme business conditions: Medical, Insurance and Airline. Due to the study's requirement for extensive data and access to leaders, these were publicly traded firms.

Although construction did not make this list, we do not advocate selling out and starting an airline. There is an adage that the best way to make a small fortune in aviation is to start with a large fortune. Instead, we are saying that construction contractors can learn from Collins and Hanson's conclusions about thriving under uncertainty and change. This is what we discuss in the remainder of this section.

DOI: 10.1201/9781003290643-8

The Concept

This book is not a continuation of Collins's previous work but one that asks a new question, "What do the superior performing companies do differently than their able competitors in a changing and risky business environment?" This research caught our attention because of construction's extreme business characteristics.

Collins, Hansen, and over 20 researchers performed the tedious work of data capture, quantitative and qualitative analysis. We should be thankful. This rigorous approach is a contrast to bestselling, but anecdotal and observational, business texts. The authors analysed extensive facts, then created theories – the only time opinion and experience were used – and tested those theories. All this was done over months until they were satisfied with the research's depth, breadth and quality.

The following sections outline the attributes and behaviours of 10x companies compared with their less-successful counterparts. First, we describe Collins and Hanson's ideas and then apply them to our industry.

Level 5 Ambition and Leadership

Level 5 Ambition is the generative mentality we want in our workforce (see our section, *arriving at a Generative Safety Culture You Will Need to Take the Winding Road*). An executive with a Level 5 Ambition feels an internalised responsibility for the firm's continued good results. This leader works selflessly but fiercely. their ambition is focused on the firm's success. Since the CEO of any organisation dramatically affects company staff by their tone, pace and practices, the method and beliefs of this person have a lasting effect on their company. Their "fingerprints" (resulting culture) have the most significant impact on the firm of any one person.

Level 5 Leadership means inspiring people from this focused but selfless perspective. These leaders' determination is about the prosperity of the firm they head. These kinds of leaders think about "who" rather than "what" in their decisions. Construction is a human-enabled industry: someone's hands have touched every put-in-place piece of work. When those hands are focused on the firm's success – and this commitment is modelled by everyone right up to the top leadership – the company can thrive in uncertainty. When a site supervisor is willing to operate equipment in a recession, this leader models a successful attitude for all employees.

SMaC

SMaC is an acronym for Specific, Methodical and Consistent. These are the core practices that the 10x companies implemented and did not change for years. The change was to only 15% of their practices on average when they did. Less-successful comparison companies modified their practices on average 60%.

We have been writing practice manuals for our clients since 2009. This is a core service that has a specific process that is a methodical and unifying approach assuring consistent results as outlined in the research.

SMaC is comprised of three principles and bounded by "Level 5 Ambition". They are 1) Fanatic Discipline, 2) Productive Paranoia and 3) Empirical Creativity. We discuss these last three concepts in subsequent sections.

Fanatic Discipline

This is a sub-attribute of SMaC. It is foundational to high performance. As contractors know, compliance to processes is the first best practice – laxness in execution will cause financial losses. Since pre-tax profit is single digits in the industry, costs can easily exceed turnover due to one project mistake. This same mistake may cause negative cash flow (unpaid progress billing), leaving the contractor financially stressed. They may have to borrow funds to pay for current obligations, increasing their operational cost.

Collins and Hansen use the example of the 20-mile march to illustrate one part of fanatic discipline. 10x companies do not extend themselves when a great opportunity presents itself, nor do they take days off when the going is rough. The concept of the 20-mile march is the same as the best craftspeople use – steady and focused progress while keeping one's eyes open for changes.

Competing Antarctic explorers Amundsen and Scott acted in contrasting ways and were used by Collins and Hanson to illustrate the 20-mile march concept. Amundsen was the first to the South Pole and returned to his country a hero, while Scott died on his return trip. Each took a different level of care with basic management tasks such as planning the journey, executing the plan, measuring progress and creating failsafe systems. Suffice it to say that Amundsen took more time with each pre-exploration day to plan each task, create redundant systems, anticipate things that could go wrong and make alternate plans in an unforeseen event.

The 20-mile march reminds us of what we see in growing but highly efficient construction contractors. Quality growth is one way to describe this method of managing effort over the long term. As construction people know, exponential growth exposes a firm to the risk of working with unknown clients, designers and vendors while employing new staff unvetted by time. We have worked with solid companies that have turned down millions of dollars in work to ensure that they grow slowly and sustainably, with integrity to their core values, safely delivering quality put-in-place work every time while making money in the process. Their current clients appreciate the focus on slow growth since they continue to receive quality construction service. We have also worked with people who just want to "get 'er done" and are willing to take on large projects and introduce unnecessary risk into their companies to chase more turnover opportunistically. It is essential to point out that the best clients and craftspeople reward construction's 20-mile marchers with loyalty and longevity.

Productive Paranoia

Productive paranoia is exactly what it states – anticipating adverse events regardless of the amount of evidence. However, the high-performing companies were not overwhelmed by the possibilities but took direct action to eliminate or minimise those potential events – the most significant threats being first on their list.

Examples include 1) Instituting a "do-over button" accessible to each attendee in (poor) pre-construction meetings. As a result, the first time a session is repeated is usually the last time it happens 2) Having a project executive hierarchical structure lessens young project manager mistakes. Everyone has a "gap of knowledge" – the project executive is a resource to fill those gaps 3) Meeting with a client face to face when there is a need to ask confronting questions. their answers and accompanying body language will tell most of the truth.

Empirical Creativity

Collins and Hanson argue that data spurs creative thought is a core practice of high-performing companies that operate in a changing and risky business environment. In any business – and especially in construction contracting – the number of different opinions is the same as the number of executives in a meeting. The start of any problem-solving discussion is not always smooth, with many approaches to defining and solving the problem being expressed. However, if data starts the discussion, any group has a credible starting point about defining and solving a problem.

In looking ahead for the next profitable insight – outpacing the competition – data is the ultimate source of facts. Contractors have found many unrealised insights about the construction business from both industry data and their own. One industry example is paying accounts early. Paying bills earlier than required is a trait of the top quartile contractors. Conversely, below-median performers have weak pre-construction planning practices or a philosophy of overhead frugality.

Fire Bullets, Then Cannonballs

The concept of "Fire Bullets, Then Cannonballs" is an instruction for incremental innovation. Bullets are small innovative actions and cannonballs are large bets on that innovation. As one can guess, bullets are relatively inexpensive and quiet, while cannonballs are not. Indeed, the market is the ultimate judge of an invention paying off. Therefore, making many small bets then waiting for the results, no matter how strongly one feels, is sure, maybe the best long-term strategy.

Given that there are many small bets, there is no predicting each success level. Certainly, there are ones that will be modest successes while others will be major ones. Those initiatives that prove to be enormously profitable are where any company will follow up with significant resources (cannonballs).

To listen to business pundits and popular authors, the outstanding impression is that innovation is a differentiator. However, high levels of innovation do not distinguish high-performing firms from their lesser competitors. A great amount of effort was spent by Collins and Hansen to find out why. The authors found that a minimum level of innovation is necessary to survive. As a result, 10xers kept one step behind their competitors' technology. There is a corollary to this idea in construction: a minimum amount of technology is required for any contractor to keep headcount down and process speed up. However, the use of technology does not ensure profits.

In construction contracts, we have found that adopting technology requires four expensive investments 1) Software purchase, 2) Employee training, 3) Customised programming (due to the unique nature of every construction firm's operation) 4) Normalisation time. This last one can take as long as 18 months for a technology conversion while realigning processes. In our 4% net profit business, if these four investments cost $200,000, then $6.6 million of extra turnover must be earned to break even on this cost. We have not included the cost of inevitable errors due to technology conversion.

An astute contractor may wonder how to fire bullets for a major initiative like software. We recommend that a company shore up its processes first, making small investments in Standard Operating Procedures. This can be a qualifying document for software proposals ("Can your software follow our process?"). Once in place, new software can be implemented that aligns with the company's procedures and the SOPs can be updated as the new software requires. Unfortunately, we have seen companies attempt to implement new software without a core set of SOPs in place. The results are often disorganised, inefficient and embarrassing.

Leading Above the Deathline

Leading above the Deathline is ultimately an awareness of the many uncontrollable and deadly factors residing in the constructing process. High performing companies are conservative in their preparations for unplanned adverse events. They hoard cash and execute their contingency plans quickly. One example cited in the book was Southwest Airlines which turned a profit in the crisis years of 2001 and 2002. They made more money than all of their competitors combined during this period. This added to their industry-leading cushion of current assets.

A construction contractor needs enough cash to meet expenses but should not hoard so much cash that they under-invest in equipment, planning or personnel.

Zoom Out, Then Zoom In

This is the authors' curt way to explain strategic planning. First, the big picture (Zooming Out) assesses the environment and then plans strategic actions (Zooming in). Collins and Hanson's study concludes that this is a consistent habit of high performers compared to their less-successful competitors.

Undoubtedly, construction firms' need to plan their strategic pathway is more significant than ever before. Zooming Out is arguably the most challenging part – what does the future hold? We would add our construction contracting-specific framework, GRAET helps strategic planning. It represents a long and efficient journey.

Logically, if lower-level employees learn this practice, then senior level people will be aided to overcome their leadership challenges. In addition, if many employees are aware of strategic planning's requirements, it leads to more insightful conversations.

We have observed companies that journeypersons, labourers and other non-management employees are encouraged to suggest new ideas for the company – and are rewarded for doing so. This has resulted in an efficiency that adds to the other previous improvements.

If business development is only an executive responsibility, many opportunities will be missed to extend this thinking. However, if employees know what a good project looks like, this gives a company more eyes on the market. By the way, the best business developers we know are "brown socks, short sleeve" types. As you can guess, the most sophisticated construction service buyers love dealing with an "operator", not a prototypical salesperson.

Risk Management

Formalised Risk Management has become a mainstream topic in the last few years. For example, it is a standard subject in graduate construction programs at major universities. In addition, banks and other financing providers companies have emphasised risk management more than ever.

One odd feature of construction contracting is that the less risky the project, the better the reward (return on investment). Even for high-risk projects, building it on paper with contingency planning has anecdotally proven to be a practical improvement in increasing reward and reducing risk. Contracting a much larger than a contractor's average project size is avoided if Risk Management is seriously practised. It is for a good reason, as cited in the *Great by Choice*. It can increase three kinds of risks – Deathline, Asymmetrical or Uncontrollable.

Not managing risk can lead to a construction firm's demise. FMI's http://www.fmiquarterly.com/index.php/2016/06/14/why-large-contractors-fail-a-fresh-perspective recent study listed root causes for significant contractor failure. Each of these is controllable: 1) Too much change – 90% 2) Poor strategic leadership – 76% 3) Excessive ego – 62% 4) Inadequate capitalisation – 58% 5) Loss of Discipline – 45% (FMI Quarterly, 2016).

Return on Luck

For anyone in business over the long term, luck will find you – good or bad. Planning with this in mind makes a company quicker to respond to adverse or advantageous situations. For the companies studied, the high performing firms executed faster and more completely when each surprise occurred. Collins and Hansen share how each industry pair studied planned differently.

Of course, construction executives should be prepared for windfalls and devastations always. That is the benefit of productive paranoia – getting an early heads-up of what might be going on – effective risk management. Managing through reverses is never an easy process; however, having a ready-made plan for a client's bankruptcy can put you at the courthouse first. For example, in the case of a voluntary employment separation, you may give an unanticipated promotion to someone well-qualified but underutilised. Even though the supervisor may say, "She is not ready" (possibly hoarding talent to make their life easier), this might be the excuse needed to promote this young leader.

If a competitor goes bankrupt, it is appropriate to contact the project owner or surety to describe your financial strength and depth of management while inquiring about a time and materials (T&M) contract on uncompleted contracts. Additionally, if you are taking over work on this basis, it might be a good time to slow bidding and refocus some work-acquisition employees on these T&M contracts. The site experience your people receive should have a long-term benefit. This is another place where it pays to have a generative climate. Most of the time, someone who does business development can also hang a door or create and update a project schedule; repurposing staff will not be hard. It will also keep them nimble and practised for future economic recessions and expansions.

Compliance to Process

Compliance to Process is the basics matter in any business. Unfortunately, much money is lost in construction by not executing simple, everyday and overlooked intelligent steps. One of our clients calls these oversights "the cookie jar of millions". In the construction industry, performing the fundamentals will continue to be an opportunity for those who understand it and a risk for those who do not. Collins and Hansen make it clear that fanatic discipline is a differentiator between those 10x better than their peers.

So, double-checking a restricted access gate, repeatedly asking about an employee's mindset or persistently chasing a Request for Information (RFI) makes it more likely that the construction contractor will avoid a significant monetary loss, morale decline or tarnished reputation. Of course, we say this differently than Collins and Hanson. However, the message is the same: compliance with fundamental processes proves a company's level of its fanatical discipline and is strongly correlated with its success.

We have found that assuring the business sustainability of a construction firm follows the same process. The first step is clarifying and documenting management practices. Coupled with a visual map and compliance measurement, this basic approach of "the one way that works" takes the ambiguity out of business, freeing up time for managers to think and lead (not firefight). It is also a training document and performance standard for millennials. Millennials want this so they know what to do to be successful (See our section, *Understanding and Managing Millennials: A Crash Course for the Uninitiated*).

Conclusion

Collins and Hansen have given the construction industry a gift. We believe the value of their research is high and should be weighted more heavily than most other business books – especially many current ones, which are anecdotal and observation-based.

We do not believe better practices are mysterious or difficult to find. On the contrary, we sense that the answer for the construction industry is in front of all of us. So, do not be distracted by the next bestseller. Watch the companies with the most efficient and durable performance, not those who do the best job promoting themselves or are the largest. This is a trap.

Indeed, those companies that are less efficient contrast the best. This paring is a joint research practice, but it has not been done in the construction industry. It would likely highlight significant strategic, cultural, and operating differences if done.

The largest lessons from *Great by Choice* are just three: 1) Fanatic Discipline, 2) Productive Paranoia and 3) Empirical Creativity. In short, executives with Level 5 Ambition are led by executives – an ambition that filters down to all company levels.

In contrast, using opinion or top-line turnover as the starting point of any discussion is deemed ineffective.

Our observation is that leading a contracting organisation is firmly rooted in being an able constructor. A construction firm's primary economic value-add is technical mastery of its trade. While good craft skill is essential, the greater challenge is leading people, managing processes and making timely decisions. And yet, the industry struggles to get these things right. Why? One reason is research: construction organisational performance appears to be one of the most understudied issues in any industry. Data availability is a significant issue: many sectors have a substantial number of publicly held firms, which leads to easily accessible information, including financial performance. The construction industry's publicly held firms are only a handful out of approximately 385,000.

Our firm has carefully compiled a set of efficiency data and continues to add to it. As a result, what started as Stevens's accepted dissertation has become a more extensive and more relevant data set. This proprietary research distils practices that efficient contractors – top half in overhead/direct cost ratio – use and continue to outpace unaware competitors performing below the median (half of all contractors).

Highly efficient firms do not want to be studied by university academics and become the subject of a case study. Instead, they desire to "make money alone in the dark". The resulting lack of industry information means that efficient practices are not widely known. However, construction contractors who pay attention to existing evidence can move ahead of their peers while experiencing a smoother business journey.

Managing Giga Projects: Advice From Those Who Have Been There, Done That

Giga projects are the most extreme projects built. Their size and complexity represent the largest of all challenges for any constructor. This book, *Managing Giga Projects: Advice from those who have been there, done that* is an omnibus of lessons learned for construction professionals. It is written by those who have the benefit of mistakes and successes. No one should wait until building the largest job of their career to read this book.

We believe the 26 authors (the editors were also authors) contained in the book capture the common struggle of "Giga" projects as 1) overcoming complexity, 2) managing man-made and natural risk and 3) being prepared for the arbitrariness of luck. Each area is not unique to Giga projects; every construction job has each in the three areas.

Readers are in for a deep and broad journey literally – unique projects from all over the world that sometimes require the development and installation of highly technical systems. Part of this journey appears to be a historical one in most chapters. Hence, the construction industry background and thinking in each area covered. *Stellar practitioners created managing Giga Project*s, financiers, owners, program managers, consultants, designers, contractors and legal counsel. Most hold senior positions in global organisations. This is a learned group that can synthesise project lessons into understandable insights. As you might guess, the book is overwhelming in its content.

Just as there is no perfect journey in one's career, this book says there is no ideal pathway to building a huge and complex project. This is the first lesson of the book. The journey cannot be prescribed; it is unscripted except for significant milestones and overall processes. There will always be uncontrollable events such as credit crises, governmental actions or latent conditions. Therefore, risk planning can never be ignored.

Editors Galloway, Nielsen and Dignum helped the quality of thought in this book. As good editors do, they challenged the authors to increase their clarity and detail orientation, producing more understandable and connected thinking.

For our industry, the book's contribution is to share explanations of and possible actions to address many typical large project and program management issues. Anyone who understands construction is improved by digesting the book's vocabulary, concepts, reasoning and conclusions. This section is our view of the book's value which we will articulate below.

Contrastingly, being prepared to take advantage of good luck is rarely talked about. Of course, if one is prepared, it represents a real opportunity to deliver the project with better safety, quality and cost while completing it on time. Taking advantage of luck acts as a kind of insurance policy for predictable reverses all Giga projects experience.

Having the discipline to plan for most contingencies and staffing experienced people to troubleshoot problems are central themes of the book. This might be called wise leadership. Unfortunately, savvy is rare in most professions. The gift of this book is that it gives insights and examples of it in numerous and varied ways. Of course, it is up to the reader to study, adapt and extend this thinking to their current situation.

There are numerous case studies – over 50 – of projects in globally diverse places. Each by itself would be enough for an individual book if expanded. Reading about the significant challenges in major projects provides a picture of sound strategic and operational practices. We assert that these lessons apply to all sizes of projects in which safety, quality, cost and schedule are the top 4 priorities.

The most compelling lesson, we think, is the focus on "perfect information". Perfect information is a Wall Street term meaning the complete collection of facts, data and hedges about a prospective investment decision. It is not written per se, but it is the largest theme. From design to pre-construction to construction and operation of a Giga project, having a complete set of facts including data and "what if" scenarios give the leadership team all possible information; from there, it is a matter of philosophy and discipline quality of the project's outcome. Furthermore, several authors sternly warned about the danger of receiving "filtered" information. We interpret this as a reference to employing "yes" people or politically sensitive employees. A take-away from this might be to hire those who deliver unembellished facts and raw observations.

Managing Giga Projects Chapters' Summaries

Each chapter is filled with summaries, details, insights and case studies from each author's experience and research. Due to the space afforded to this review, our outline describes our interpretation of the major themes of each.

Chapter 1 deals with governance issues, including managing liabilities, communication, expectations and the overall program. Nielsen's first piece of advice is deft: Manage project expectations because that becomes other's measuring stick of project performance. The last part of this chapter outlines what kind of policies, procedures and processes should formalise the overall governance of the project.

Chapter 2 addresses risk management practices involving large projects. Nielsen, Dignum and Reilly outline many large areas that contain most of the risk, such as delivery, operational, technological, financial, procurement, political, environmental, social and economic. These are broken down into smaller subcategories and discussed in detail. A risk checklist and quantitative assessment system are included in this chapter.

Chapter 3 discusses financial markets and their effects on projects. Malek addresses this subject from the energy sector. He uses the 2008 Great Financial Crisis as an example of the worst-case real-life scenario. It is an appropriate before and after picture. Malek points out the following lessons for contractors:

- Construction executives are the most extraordinary business optimists, sometimes unduly.
- The deeper the niche a contractor masters, the more safety net their business turnover will have.
- There continues to be a lack of contractor operational transparency. Owners often do not know how busy a firm is and the bottom-line status of a specific project?

Chapter 4 discusses project delivery methods. A myriad of ways is listed; however, Hughes notes that there are only a few basic ones that add layers of management and control. The need to ensure value-for-money competition and transparency is a significant consideration in determining a method. He cautions that all delivery methods create tensions between parties. Sensitivity to this minimises unnecessary pressure. Of high importance, the owner must be protected during and after construction and many times; the public is a vested stakeholder.

Chapter 5 addresses public–private partnership contracting and discusses its applicability to Giga projects. Little suggests that all public projects are political and Giga ones are more so. Their cost and time results are much worse than initially budgeted and presented to taxpayers. He contrasts this with the demonstrably better performance of privately funded and managed large projects. In this way, he suggests logically that the injection of private ownership and its inherent cost/time discipline has value.

In *Chapter 6*, Prieto fully describes the program manager's role. It thoroughly explores all possible considerations in framing and executing their role, responsibility and authority. It is a robust list of references that only the experienced executive knows. What is interesting is that program managers must be able to create unique solutions. This is akin to NASA engineers who must plan on systems and materials not yet invented to reach their goals. Conventional ideas cannot be the limit of their thinking.

Chapter 7 deals with financing megaprojects. Tucker asserts that financial success is determined by whether the initial published cost estimate is met. This is one of the two largest expectations set (schedule) that can doom perceived industry and public success. Furthermore, financing is the lifeblood of a project. This is a seminal event of "go or no go". There is more caution in approving this critical component due to the financial history of large projects. There are many examples of cost overruns after initial budgets were approved and it has transformed owners, users and taxpayers into sceptics.

Chapter 8 outlines six challenges to controlling projects. Finally, for those who like to skip to different areas, Galloway and Reilly articulate six core strategic areas for Pre-Design planning:

1 The ricochet effect.
2 Controlling non-participatory stakeholder's expectations.
3 Controlling cultural differences.
4 Controlling cost creep.
5 Controlling schedule creep.
6 Controlling information overload.

These are the largest management components of a project, whether it is Giga or not. Indeed, due to the formation of a team of thousands of people, hundreds of decisions on any day lead to these potential problems.

Chapter 9 explores the issues of managing the design of Giga projects. Warne walks the reader through the most significant issues of designing this type of construction. Getting the constructor involved early, although commonly recommended now, is critical to give the project a chance at a fast start. A contractor's familiarity with standard installation sequences and details, which the design might accommodate, can smooth the construction process. Additionally, having iterative and consistent design reviews can take a good design and make it exceptional, giving more consideration to details which can speed up and make installations safer. He also addresses post-design (while construction is ongoing) issues and cautions Giga project owners not to scale back design staff too quickly due to the continuing support needed to clarify further and troubleshoot construction details on site.

Chapter 10 addresses procurement and construction management. Crumm details the major cost areas of all Giga projects: procurement of project staff, material, capital equipment and construction service providers (subcontractors). In their view, this area covers up to 85% of the total project expense. He notes that planning the execution (not the schedule) is the critical piece that is sometimes not created with the care or followed when events inevitably occur.

Chapter 11 discusses the culture of corruption and ethics in international projects. This is a significant waste of capital and goodwill in projects that ethical leaders do not govern, whether they are construction, regulatory, governmental or societal. Henry discusses the roots and effects of corruption and what can be done about it.

Chapter 12 encompasses dispute resolution by Hinchey. This is a critical factor to keep the project from being delayed, parties distracted and the project in the public eye well after completion. Many forms of dispute resolution are discussed and guidance is given as to how to proceed. As the author recommends, early identification and processing of disputes help avoid unnecessary delays. However, standardisation is not yet part of this area of practice, so care must be taken to understand components such as local jurisdiction and practices.

Chapter 13 to *18* contains several authors' case studies and their analysis of projects in the Middle East, U.K., South America, Asia, Australia and North America. These project summaries include country idiosyncrasies and future trends. Due to the limitations of the space constraints of this explanation, high-level facts and charts are given. Insightful observations should help the new practitioner to that country cited.

Chapter 19 addresses the challenges of the most extraordinary Giga project: nuclear construction. Whitney. Bloodworth and Sanacory describe and explain the complexity with which these projects are planned, designed, built and turned over to the owner and its operator. They present all the risks that flow from the first day.

More nuclear facilities are being built outside North America than in it. 58 by the author's count, of which 20 are in China. There is new base-load demand in growing countries. They are not opting for coal-fired plants has facilitated the demand for nuclear energy. Green energy in its present technological state cannot supply these increasing needs.

Conclusion

There are only a few books that apply directly to the construction industry and, at the same time, are a summary of multiple accomplished professionals' experiences.

One of the great values in this book is that there is less bias or "tunnel vision" due to its many authors and trio of editors. We cannot think of another work with a more varied and experienced perspective of the largest construction projects. Further adding to its value are the discussions of Giga project challenges such as technology innovation, politically sensitivity and human diversity.

This book illuminates the complexity and other major challenges facing any construction team for the outsider to our industry. One insight is the unpredictability of an adverse event's "ricochet effect" – a small, seemingly insignificant occurrence can have large and lasting impacts on a Giga project. The experienced and brilliant team of constructors can stay ahead and manage it. Unfortunately, these kinds of groups are rare.

One Giga project philosophy is the idea of the 100-year view. The project will outlive its stakeholders, so the long-term values of managing expectations, planning, discipline, transparency, communication and collaboration are essential. We think this list of management challenges is equal to the task of determining technical details. They are that important. The book's authors point out several times that choosing the management approach for the project is critical for it to have any chance of economic and technical success.

We suggest that understanding any deep, broad and complex book fully requires re-reading. This is one of those works. However, once understood, the education from this content should help the reader be more aware, i.e., fewer surprises and more potential answers to any project challenges, Giga or not.

Freakonomics

Authors Levitt and Dubner state that there is no unifying theme in Freakonomics. They believe there is not a common thread of thought. Consider. There is one and such diverse organisations practice it as Morgan Stanley, The Government and successful construction contractors. What is the theme? *"Perfect Information"*.

Perfect information is an asset that is invaluable to any investor, double agent or construction executive. As you know, the world is not static. It changes for better or worse each day. All changes affect any asset. With perfect information, you understand how and when to act. Do not doubt it; the practitioners have proven this concept over decades, especially successful construction firms.

Acquiring perfect information gives you two valuable options:

1 *Face the challenge and win.* Any person or organisation can win more often through better insight about the opportunity than the competition.
2 *Do not participate and let the competition struggle with the opportunity.* Since you know it is negative, you walk away and spend your resources on something else worthwhile. Meantime, your peer is distracted and occupied with a problem(s). The poker world calls this "getting out of the way" of others. Let competitors who do not know all the facts become weaker by their own actions.

How to get perfect information? There is a process. The simplest way is to acquire trustworthy data, follow up with your human contacts to clarify and synthesise the data into an actionable plan.

In the PhD research world, the process is as follows:

- *Literature review* – reading many worthwhile research papers gives you an understanding of the level and reach of current research. Peer-reviewed journals, quality periodicals and the like will provide you with many ideas along with the current status of your particular interest. For contractors, several industry and government sources are also available.
- *Data Collection / Quantitative Analysis* – create the hypothesis, structure the research process, solicit qualified participants, received the data and analyse it. Objective measurements tell an accurate story. It limits bias. As you know, we have many in construction.
- *Qualitative Follow Up* – ask the humans involved. Have questions ready for them from your data analysis. Collect all observations and opinions from the knowledgeable and experienced people who know the background, causes and effects of your area of research
- *Synthesise* – makes sense of it all. Use frameworks, processes and intellectual power to see what conclusions are valid. You must determine what the information is telling you. This is the hardest part. It takes iterative and intermittent days to tease out complete understanding.

Levitt went through all these steps and made several conclusions that are fact-based. His books are a product of that. This is no small feat of energy and insight.

The author shows us that questions are never old, but data does become old. He reads many things that interest him, asks questions, seeks data, and then follows up with people who know that area well. From here, his intriguing mind sees patterns and aims to confirm guesses. Finally, the book is a summary of this methodology reaching conclusions.

For construction firms, perfect information can be distilled by using an OATA Model:

Observing – see data, read current information and watch trends. We must acquire quantitative information or statistics. Good data is fact-based, or as we have stated, "it is not an opinion". It tells you points a direction and is trustworthy.

Asking – ask why, how, what, when, where and who? Data and people will give you answers. Ask those who do, those who manage and those who lead, or the numbers before you. A day spent here can lead you to a cost reduction or a turnover enhancing idea by asking questions prompted by the above statistical analysis. Unfortunately, this anecdotal and qualitative information is not readily available and can only be found through extensive research (or asking).

Thinking – What does this say and how does it apply to our construction business? How does this information affect profit or cash flow (or both)?

Acting – Taking a direction and committing resources. Considering reasons slowly but acting fast when conclusions are made can be game-changers. Indeed, only a few ideas will be worthy of this last step. Leadership requires bravery. Bravery is a function of a well-reasoned direction.

For construction firms, executives cannot be doers only; they must be thinkers. Therefore, knowing all the information helps make quality decisions about strategic direction and subsequent resource allocation. From this point, middle managers can efficiently implement the direction by keeping costs low and turnover collected.

Serious contractors who see their businesses as an opportunity to increase operational return on investment and organisational value practice the above two.

This lesson must be known for those we are training to be our replacement. We think it is a litmus test for anyone to earn a leadership position.

Levitt and Dunbar cite specific data in this book and interpret it well. It is an interesting journey of eclectic facts, although some intellectual rigour may be missing. Here are some examples; these are their chapter titles.

1   What do School Teachers and Sumo Wrestlers have in Common?
2   How is the Klu Klux Klan like a group of Real Estate Agents?
3   Why do Drug Dealers still live with their Moms?

These questions and more are answered systematically. There are no quick, cute conclusions. The logic and insight are compelling.

Applying Freakonomics to Construction Contracting

As we have stated, insights can be derived from statistical data. The Construction Financial Management Association (CFMA) is a stellar information resource. It publishes several insightful guides focused squarely on construction contracting. One publication in particular, "The Construction Industry Annual Financial Survey", is our industry's relatively large data set. Statistically, it provides credible evidence of the efficacy of several financial practices of the most profitable contractors. It is exciting reading – 300+ pages – to strengthen your profit & loss, balance sheet, and cash flow statement. If you are serious about your financial management, it is required reading. Additionally, it is even more critical if you are worried about your retirement.

As an example of insights gleaned from the data contained in the research, let us share three metrics of financially successful contracting firms. That is, those in the top 25%of return-on-asset or return-on-equity metrics. These may surprise you:

1   Turnover per contract – less than the average firm.
2   Average working capital turnover – slower than the average firm.
3   Fixed asset ratio – less fixed assets than the average firm.

There are many more. These and other characteristics tend to be counterintuitive in our industry. Some of these assertions defy financial common sense.

From another source, current banking data tells us about other unusual statistics regarding active construction firms. For example, it indicates that the highest quartile of profitable contractors pays their bills 20% to 45% faster (in days) than the median quartile. This is the most substantial evidence proven in the best data for the construction contracting industry.

Let us assume that your competitors pay suppliers in 46 days. Let us also say your policy for payment of billing for completed work is fewer than 46 days, say 36 days, or 10 days earlier than the average. This equates to .2% interest. Our math is as follows:

assume 8% interest per annum divided by 365 days. This equals 2/100's of 1% per day. Multiply this by 10 days results in 2/10 of 1% interest lost by paying early.

So, ask yourself: Are there any benefits described above that you gain from paying just 10 days earlier than average, worth more than 1/5 of 1%?

Certainly, several come to mind:

- Faster turnaround on quotes/bids from others.
- Quicker attention on deliveries or shipping.
- Some flexibility on pricing from vendors and service providers.
- Faster resolution of conflicts or disputes.
- Better response to manning a project.
- Friendlier terms and conditions of sale.

Is this worth 1/5 of 1% of cost? We think the answer is obvious in our 4% net profit before tax business.

Reinvesting profits takes on a new direction. Never mind the new computer system or office addition. Instead, think deeply about increasing your working capital and, thus, your ability to pay vendors and service providers somewhat faster while asking for more generous terms, conditions and service.

So, as we apply the lessons of Freakonomics to our everyday construction challenges, one of the gifts of Levitt and Dubner is that we can outsmart our competition if we are willing to observe, ask, think and act.

Remember, people who know everything about construction contracting will eventually have outdated knowledge. Change is constant and you must catch the difference. Be observant and pay attention to current data and observable trends. Tomorrow is a new opportunity.

Since change is consistent, the only option is to work with it. So, asking how to benefit from a change will keep you ahead of most competitors. Some people resist change. That is human and rational, but a competitive edge for you.

Knowing more than your peers empowers you. You can dominate a situation and win. The construction industry rewards participating in good situations. However, the reverse is also true. In poor business situations, one must simply get out of the way. The common denominator is to acquire superior information and act on it. It is a skill that all executives need in any economy.

Mastering the Rockefeller Habits

Verne Harnish's *Mastering the Rockefeller Habits: What You Must Do to Increase the Value of Your Growing Firm* makes the challenge of construction contracting management easier. A rarity in the annals of management thinking, it includes eight time-tested, practical processes adapted from some of the best-run firms that executives can implement to strengthen their businesses. The principles discussed apply in whatever location or language the firm is operating. It should be the opening text of any leadership or people management course.

Some of the leadership recommendations Harnish discusses:

> Focus on the "what", not the "who." The first question to ask is "what is the right thing to do?" not "whom it will upset" or "who gets the credit?"

Behave well and be polite regardless of others' actions. But behaving in a mannerly fashion does not banish the words *"no"* or *"pass"*, or phrases such as *"that's harsh"* or *"I would like to see the data"*.

The strategy must be clear enough to be written on one page. Clarity and conciseness are the hard work of strategic planning. Harnish shares examples of a single page strategic plan. Brevity positively impacts understanding. Having understood at all levels of a firm leads to purposeful collective actions.

Laziness and leadership are negatively correlated. Being over-prepared is a characteristic of a serious leader. To lead a substantial construction organisation, read many books, facilitate several efforts and interact with dozens of professionals gives perspective. It creates a sensitivity to people that helps in several ways. Experience is a valuable commodity. It is a function of time, effort and focuses on the task. Young people, simply put, have not existed long enough to have significant experience. When they have exposure in a multifaceted way, it builds leadership skills over time.

This book should be part of any college-level course in construction labour management. Since it takes many years to manage labour well, so starting in college makes sense. In addition, mastering *the Rockefeller Habits* illustrates confidence in a method that most young supervisors can adopt immediately.

Human Resource Management

Managing people is inefficient because people are not 100% reliable. They sometimes work with less than total quality in mind. Regardless, they are critical to the construction industry. Human resource management is *the* leverage point in all construction. Done well, the frequency of daily crises lessens and projects are built smoothly. Quality productivity is consistently increased by improved people management.

Knowledge about people can never be complete. How to manage them perfectly is a worthy goal but only an ideal that can never be reached. Unlike Swiss watches, people have too many individual differences – even those born from the same parents and raised expect that any two can perform the same task precisely. But it is both a blessing and a curse that no two people are the same.

Harnish implies that good parenting and effective executive management are closely related. Raising children to be dependable, independent adults is the primary function of good parenting. Likewise, raising inexperienced employees into reliable, independent workers is critical to achieving a smooth operating construction firm.

Harnish recommends having a few critical rules. For example, attention to detail, cost discipline, daily planning, resource forecasting and truthful scheduling are vital for a construction firm. Indeed, there are others.

Repeat your essential rules often and in different ways. This helps to keep people reminded but not annoyed. It is a skill that teachers and executives must acquire to be successful.

Then live the essential rules and teach by way of example. This is a powerful way to keep your small number of principles alive and clear to your subordinates.

Few professionals are "plug and play". New hires to your firm will go through a "norming" process in company culture, politics and procedures to be optimally productive. Therefore, a CEO must be a master at (or delegating) getting the potential out of new entrants.

Improving individuals in your workforce – even those who spend their entire career in the same position – is a worthwhile and practical goal, essential for the sustainability of any construction organisation. Therefore, one of the primary duties of the CEO is to make sure the firm has bench strength in professionals who can contribute at the next level. As chaos is inevitable in the construction industry, this lessens the risk of a disloyal workforce (those who have loyalty to themselves but not to the companies they work with).

In teaching employees, Harnish recommends that trainers be sensitive to people's learning styles. Part of leadership is communicating efficacy (effectiveness and efficiency) to those who follow. This is difficult work and worth the effort. A lack of understanding by employees leads to weak leadership regardless of the cause.

Meeting Management

As much as we wish meetings would just go away in the construction industry, they never will. When conducted well, meetings are opportunities to engage others, direct thinking and confirm agreement. In our opinion, they are an efficient way to gather information. As has been said by many others, people communicate most of their messages by non-verbal clues.

Learning to function well in corporate meetings will earn you respect from your peers, subordinates and clients. In other words, your professional standing will increase. Each of us has spent more than once in unproductive meetings. We appreciate those who lead discussions efficiently.

Harnish shares effective meeting strategies, including rhythm, timing, topics and methodology to get the most out of them, whether internal or external to your firm. He has summarised this information from many stellar firms. Given the success of the featured companies, these could be considered meeting management "good operating practices".

The Role of Data in Excellent Management

Without data, there can be no fact-based discussion. It would be all impression and opinion with each manager giving an incomplete understanding of the current situation and what to do about it. Data completes any situational picture as it provides a solid starting point and disciplines discussion. For example, risk in the construction industry is at an all-time high in the opinion of many. Making sure that reality is understood before acting is critical. Good data furnish that.

What is the data you should collect and how should it be handled? There are no standard accounting processes (Generally Accepted Accounting Principles) for contracting. Many families own construction firms where less rigorous data collection practices are standard. However, several stellar data sources, including some provided by the government, are available for firms to compare their operations to their peers. Some caution is advised here: the variables applicable to your firm may not precisely compare your operations to the statistics provided in the general charts. Overall, the resulting lines tell a story of a weakening or strengthening organisation when data is trended. This is a better way to treat most data.

Financial Matters

A personal relationship with financial industry professionals is critical in all businesses. It is crucial to have one, so it should not be ignored. Construction is no different. Banks, sureties and insurance firms hold necessary purse strings, so engage these persons from the start of your firm's life. Contractors might be generally self-reliant, but they will not be so continuously. Business reverses can and will occur. A good relationship with the financial industry offers many benefits, including a buffer for these problematic events. Furthermore, any significant project award travels through banking, insurance and surety qualification. Your financial partners can help acquire future work.

Substantial contractors work closely with their bankers, surety producers and insurance agents. To develop a strong relationship, it is essential to communicate fully. This includes alerting them to any changes in your firm and its forecast. Good firms do not shade or conceal the truth, unlike companies who get into financial difficulties. But, on the other hand, financial providers treat firms with "dodgy" practices harshly.

Solid firms in good financial shape are not afraid to communicate with outside financial parties and confirm the facts about their organisation. They lay out the past year's metrics and next year's plan. The data they share are well-vetted and they demonstrate a keen understanding of operations, including all financial aspects. This instils a sense of confidence.

All in all, a conservative financial approach lessens company and project risk. Successful construction executives take no large gambles, only well thought out directions. This makes everyone relax and look objectively at issues.

Some Areas of the Book not Aligned With the Construction Industry

Branding is an excellent consumer or business idea. It does have its place in the commercial world. However, construction contracting is not a product or just a service, but a *customised service* for clients who make their most significant capital expenditure while seeking to have the service perform quickly. In some ways, branding in construction is the quality of your last project. It is a more complex dynamic than Harnish's book illustrates.

There is an excessive focus on growth. Harnish emphasises the need for growing a firm, which is not always a good thing in contracting. Higher volume means new clients to deal with and new employees to train. In addition, new geographic areas can mean dealing with suppliers and inspection departments with which the contractor is unfamiliar. Our observation has been that these usually negatively affect profit margins.

Companies that have a unique product or service that the consumer wants can charge more since they have no competition and thus have higher profits. This, in turn, funds growth. These are often fixed-cost businesses where most of the assets are already purchased. The operating expenses (such as salaries, wages, fixed assets and materials.) are a minor percentage of the cost structure. The product sells itself in some cases. So, growth is more accessible and less risky. This scenario is unlike construction, where the craft persons and managers are the product and subject to all the inefficiencies of a custom project in a unique place. If they are new (lack experience) with the firm, this leads to more inefficiency and increased risk.

For a construction firm, it is challenging to grow the turnover line and the return-on-investment percentage.

A Synergistic Effect

Normalising your company's rules and practices into culture has a synergistic effect. After a few years of routines, the older employees will tell and show the younger employees. The tenured staff will become de facto trainers continuing the culture and transferring important knowledge. This means the executive will not. their time will be spent on other key duties.

The multifaceted Rockefeller habits have a positive synergistic effect when taken as a system. Consultants may offer a single approach and some construction professionals want "the secret". However, in construction, there is no single approach, no one secret to success. There are numerous risk factors in construction contracting. Since contractor failure is multi-factorial, it is logical to conclude that sustaining a successful construction organisation is also multi-factorial. Therefore, an organised, multi-pronged approach is needed. Harnish puts forth just such an approach.

The 7 Habits of Highly Effective People

The 7 habits of Highly Effective People by Stephen Covey gives readers explicit instruction on attaining, maintaining and continuously improving an effective mindset. From that mindset, one's actions follow. Moreover, that person can spread this outlook to peers, subordinates and even superiors. Thus, the phrase "effective people" can be used both for leaders and managers. Anyone who interacts with others benefits when understanding this book's teachings and applying them. As many have said before, any good concept only has value if applied.

This is a practical, not idealistic work. Dr Covey gathered the information for this book through his management advising work, university teaching and life experience. Dr Covey captured and interpreted their research in the real-world laboratory called "life" in most instances.

The book proposes that the primary step is for the reader to set a correct internal makeup correctly. This also can be interpreted as eliminating character flaw(s) to create a mindset more in keeping with long-term value. From the inside out, the author proposes that the reader seek to improve effectiveness. He uses several examples and even reaches back to the 1700s to explain the difference between the character ethic (then) and the personality ethic (now).

Covey purposely places the word "habits" in the book's title as he proposes that starting the journey of effectiveness begins with habits. Habits, once established, build a life of better actions and outcomes. For example, the seventh habit of improvement (sharpening the saw) delivers a stagnation antidote for professionals. My experience is that professional stagnation is an issue to overcome in business life. He addresses it directly on point.

Covey's writings contain a religious tone, but he does not propose religion as the answer. Instead, he notes that many "enduring religious faiths" have common, powerful teachings on fairness, quality and service, among other things. The book also includes quotes and examples from non-religious sources.

Covey suggests that a slow process and a self-aware mind are needed to establish long-term character. Said differently, the superior character is built carefully and methodically.

The seven habits and our interpretation for the construction industry are as follows:

* Be Proactive
* We can take control of our circumstances or be controlled by them. Many examples the author shares portray the power of "being" versus waiting to "have". It cannot be argued that the controllable part of life lies within us and the uncontrollable part outside of us. The construction contractor should focus less on the weather and government actions (uncontrollable) and more on knowing costs, planning in great detail and the like (controllable).
* Covey introduces the concept of private victories before public victories. Private victories are essential to building the confidence of the young constructor. Indeed, those who have had a reversal (such as financial, marital or professional) in their lives should be mindful of the author's private/public victory sequence. Their boss and mentor should also be. Smaller projects and less complex organisations give a manager a less intense environment to overcome small challenges. Again, self-confidence will grow.

For the estimator, proposing and pricing a variation is much easier than bidding an entire project. This is because variation pricing allows for high mental focus and a private victory. This is an example of an easy win situation that can grow a young person's self-perception of worth.

1 Begin with the End in Mind

   * The author suggests that the goal should be in one's mind before starting any personal or professional journey. Covey shares scenarios of thinking of the result first. It seems to be a large part of his leadership thinking. Quite compelling. Perhaps each construction company leader should solve the financial equation first as that is the only barrier to bankruptcy and losing their company. From there, other goals can be determined.

2 Put First Things First

   * This is defined as identifying your current centre, realising it and then moving your mindset to appropriate centres for your situation. Covey lists seven centres. He implies that the reader should not have one centre but several centres as life is a balance. If one has one centre to exclude all others, such as work or pleasure, then Covey suggests a serious re-evaluation of the person's focus.

From a centred awareness, the reader is encouraged to uncover how they spend their time and effort. An outline, descriptors and construction examples are furnished:

1 Quadrant I – Urgent and important

   * Accident on a job site

- Time-related events such as a project deadline or a lien notice rights expiration

2   Quadrant II – Not urgent and important

- Strategic planning
- New client discovery and relationship building

3   Quadrant III – Urgent and not important

- Supplier salesperson interruption
- Poorly managed meetings (project or client)

4   Quadrant IV – Not urgent and not important

- Employee birthday parties
- Donations of time and material to charities.

Starting small and planning a day, then a week is encouraged. Next, a month and a year are implied after the habit is firmly established.

1   Implementing such a system is where the value is. Using delegation is emphasised. Covey explains the two methods: 1) "Gofer" 2) "Stewardship". Delegation allows for robust production (P). Learning how to delegate is a production capability (PC) skill. All managers and leaders must note that managing things efficiently is possible, but managing people efficiently is impossible. This is important in construction contracting as people (craft, supervisors) are the means of production.
2   Balancing production with production capability during a career allows for a more substantial contribution, whatever the endeavour.

Covey introduces in this section the idea of an emotional bank account. He suggests that each of us during a relationship makes deposits (selfless gestures and thoughtful acts towards the other person) and withdrawals (asking for a favour or placing someone in an awkward situation). The sum of these two things makes our balance either negative or positive. Consistently overdrawing your emotional bank account with other shows in the other person's action towards you (i.e., lack of enthusiasm towards you or tardiness in completing your requests). A positive balance gives people "emotion air" or a good feeling towards you. Being sensitive to this account balance allows a person to manage relationships better.

Think Win/Win

- For a long-term relationship, each professional think must win/win. This ensures that each interpersonal transaction will be repeated with enthusiasm. Covey points out an attitude of win/win or no deal is also valuable. However, it is not always followed by every shareholder in the construction industry. In our experience, our industry suffers from an oversupply of contractors. Buyers of construction services know this. Specifically, a small percentage of developers seek to work with undercapitalised contractors. These buyers of construction attempt to do so to their advantage. Developers know that since these construction firms do not have a capital reserve, significant leverage may be used for mistakes or delays caused by others. With 10% retention being withheld during construction, the next payroll and this month's accounts payable is the contractor's focus. The win/win attitude should be evident in all construction business situations, but it is not.

Again, an oversupply of contractors gives some buyers no real consequence due to their actions. The next contractor is just a phone call away.

- The other scenarios of a) win/lose, b) lose/win or c) lose/lose will undoubtedly lead to no future transaction by the parties. Each of these outcomes poisons the well of goodwill between the two individuals or companies. Our industry is well served when the contracting community alerts us about abusive buyers of construction services.

Seek to Understand Then to be Understood

- This is Covey's most robust human relationship concept. In our experience, most people have been involved in superior and inferior interactions with others. We know the difference; however, the author writes several passages narrating the differences and giving examples of the most successful interactions. Whether negotiating a contract or coaching an employee, using this approach requires one first to do homework on the other person and their goals. Afterwards, a well-informed discussion or proposal may be pursued. That discussion or proposal is more likely to be trusted by the other party.
- This is a further explanation of the author's withdrawal/deposit analogy. When we seek to understand first, we are showing interest in the other person and their wants and needs. The sincerity of interest makes a deposit. We should be able to make a withdrawal later. Sadly, the author notes that there is little in the way of listening training offered in the general business world.
- Self-maturity and self-discipline are the foundations of good relationships with others. Whether we are an owner, funder, contractor or supplier, we work with some people consistently and not with others. Our experience is that it is due primarily to the other person's quality. Financial and operational issues notwithstanding, we find ways to work with those we trust. Over a 40-year construction contracting career, one's sensitivity is usually heightened. For the less experienced, this book allows an acceleration of that learning curve.

The three keys to seeking to understand are:

- Ethos – personal creditability.
- Pathos – empathy.
- Logos – logic.

Each effective presentation contains all three. They draw on the other side. Professionals who lack one of these three create a weaker connection with others than otherwise. Covey presents several examples to illustrate that is the real world.

Synergise

• The author's explanation of this is practical. He asserts that if we consider the other side and our own interests, we should be able to build more long-term value into any agreement.

Not having resistance but the trust of the other parties allows for discussions and proposals to consider more options. If we can think about what is possible, we can include more things in our agreement that will enhance value to both parties. It is simple but powerful. Trust allows us to speak frankly and openly to the other side. For example, suppose each party values the win/win concept. In that case, obstacles such as price in a construction contract can be compensated by other valuable considerations such as cash flow, future marketing opportunities and safety cost savings.

Sharpen the Saw

• Continuing improvement is a constant theme in many management texts, but Covey makes it more tangible in the reader's mind. A saw and the act of sharpening are real images. We use the phrase "competing with yourself" as a guide for professional improvement. The book uses "sharpen the saw". Ironically, this is closely aligned with the thinking and vocabulary of the construction industry. This indirectly teaches not to let the construction industry wear you down emotionally.
• The four areas for sharpening the saw are:

1 *Mental* – keeping the mind aware of what is possible and what changing.
2 *Social/Emotional* – doing those things that keep perspective in our lives.
3 *Spiritual* – the most important thing is not to work.
4 *Physical* – dedicating a part of our day to better health to keep the other three areas from being affected.

It is important to note that following this book's advice takes tremendous energy, determination and faith. It is not for the "quick-fix" minded souls of the world. Covey describes in the first chapter. It is for those who are mindful that character matters and more importantly lasts. It can be carried from generation to generation; he describes it as roots and wings.

The author has followed his advice. As a result, the book is a detailed and tightly narrated body of work that could only be produced by a focused and caring professional.

Furthermore, this book directly applies to the construction industry. It is not difficult to find leaders of construction contracting firms who are great people – exhibiting all seven habits. It is no accident that they have built significant construction firms.

Covey has built this book from a solid personal foundation. He has put great energy into this book to explain, give examples, create exercises and produce graphics. The book takes an unambiguous direction and provides specific advice for the reader. His work is remarkable and still very much applicable to today's construction industry.

The Art of War

Construction contract negotiations and the resulting operating interactions between construction service buyers (Developers or Owners), which we will label CSBs and contractors (General or Specialty) are frequently likened to battles and the construction business to war. Most construction stakeholders will understand this sentiment and see it reflected in how they must conduct business, from the proposing/ bidding stage to construction operations, including variations and, finally, project closeout. For a few decades, many have been around when CSBs and contractors worked together to create a working relationship rooted in collaboration and mutual benefit, rather than conflict. This section is for both CSBs and construction contractors and is about applying the lessons of the Art of War to create cooperation and collaboration between them.

In this age of conflict, it is sometimes easy to forget that war veterans advocate first to avoid violent confrontation. *The Art of War* by Sun Tzu Translated by Thomas Cleary, is a prominent example of this sentiment. Cleary interprets Sun Tzu's *The Art of War* from a business perspective, describing how the centuries of distilled military wisdom in Sun Tzu's volume can be applied to people conducting business in the contemporary world. We believe Tzu's lessons also apply substantially to construction. In our view, three major themes apply directly to our industry, which are all geared strongly towards minimising adversarial circumstances. They are:

1   Conflict Management.
2   Strategic Planning.
3   Risk Management.

First, we provide an overview of Cleary's interpretations of Sun Tzu's original text. Then, we list dozens of Tzu's principles, which other ancient Chinese leaders echo in different words. These other leaders are also prominent such as Cao, Meng Shi, Li Quan and Jia Ling. They are quoted in the book's pages. The efficacy of these principles was demonstrated in times of war and peace. Sun Tzu is referred to as Master Tzu out of great respect. He earned this respect through numerous contributions to his country's thinking, both in war and in peace. His advice is considered to be timeless by many leaders in both the military and business world.

Cleary's book is segmented into 13 chapters. Each chapter describes a war issue, such as strategically assessing the situation and considering whether doing battle is advantageous, how and when to apply force, take advantage of terrain and use spies. As business themes, Sun Tzu's lessons dovetail nicely with ideas of conflict management, strategic thinking and risk management approaches.

Master Tzu's foremost lesson is that risk management and conflict avoidance comprise the best initial strategy. Construction contracting is not a matter of life or death, but the daily decisions that construction contractors face can produce severe and existential consequences for the firm and its people. Below, we describe how one might apply these lessons to our construction industry, seeking first to avoid conflict and risk. The construction industry has plenty of both. Learning how to manage conflict and risk is time well spent. We can never know enough about preventing conflict and escalation.

The Art of War, then, is first an Art of Prevention. For CSBs and contractors, Sun Tzu's ideas and Cleary's interpretations for business offer many lessons. Below are a few of our applications of Master Tzu's ideas to the construction industry. Conflict and risk management and strategic planning are core aspects in each of these areas, which apply equally to CSBs and contractors:

Prevention of conflict escalation is the most efficient use of time, energy and dollars. Smart CSBs and contractors know that engaging in high-risk, conflict-escalating practices such as making up profit on variations might work as a short-term strategy. In the long term, however, he knows that what makes a firm sustainable and profitable is delivering superior work while being predictable, reliable and flexible. Similarly, an owner worth working for will pay for good craft skills, understand the process's value and be predictable, reliable and flexible. The best situation for CSBs and contractors occurs when there is a precise alignment of expectations from the beginning. Discussing potential conflicts openly in the negotiation phase is the best way to avoid escalation. Good CSBs and contractors plan to avoid conflict.

Select your counterpart whose preference is to avert a crisis. Good contractors know the best clients are the ones that are willing to discuss potential problems in the negotiation phase and are eager to work with the contractor for a mutually acceptable outcome whenever strain arises. Good CSBs know that the best way to receive excellent work outcomes is to build solid relationships and support the contractor in achieving their project productivity goals. Seeking partners in the industry who will work together in this way is advantageous. This means choosing counterparts committed to quality and productivity and willing to invest the time and resources required to put both in place.

Assist your counterpart in knowing you and how you work. Knowing yourself is a great asset. Being frank about your strengths and weaknesses helps to prevent grave mistakes. Focusing on projects, markets and customers where the company can win is a sound strategy in construction. However modest the markets, customers or projects, winning means profits; losing could lead to bankruptcy. For contractors, it is essential to commit to working the company can build well and contract with customers whose objectives and priorities concord with yours. For CSBs, this means being willing to pay what quality work is worth, accepting appropriate responsibility for changes and unforeseen circumstances and being ready to discuss productivity as part of the cost package for a job.

When conflict does escalate, do not cause your counterpart to become desperate. Despite the best intentions of both parties, sometimes conflicts do escalate in construction. Desperation causes extreme behaviour in adversaries who could significantly injure your company. Master Tzu suggests allowing the trapped, adversarial battalion a retreat path. In construction contracts, the equivalent to this would be to seek solutions that generate desirable outcomes for both parties. The best way to do this is to not see your adversary as an adversary but instead as a partner in the project. A contractor, for instance, might see an owner with a significant design flaw requiring major rework as a trapped battalion seeking a path that avoids a catastrophic loss. Give him the path of retreat. For the owner, this might mean paying for work completed, even if there is a significant negotiation overpayment for the variation that is yet unresolved. Avoiding desperation allows for a more certain future

Having highly energised enemies (competitors) who want to destroy you is a risk and conflict issue. Do not give people reasons to plot your demise actively.

Focus on compliance with good practices. Master Tzu observed warns us against the dangers of taking success for granted. Contractors who are doing well sometimes forget to comply with the good practices that put them in a solid strategic position. Contractors who are struggling may sometimes feel compelled to cut corners on compliance to move the job along faster. CSBs in a hurry may assume good build quality comes effortlessly to the contractor and can push for faster work at the expense (often unknowingly) of compliance. All of these can result in conflict. Good CSBs and contractors agree that quality work takes time. Make sure a focus on adherence to processes is a feature of every project plan and team meeting.

Support your counterpart to succeed. The ultimate goal that both CSBs and contractors want is high-quality and efficient put-in-place work. Conflict can undermine those goals for one or both parties. Master Tzu teaches that you can assist an adversary in becoming weaker if you understand him well. This implies that you can also help him in becoming more robust. In conflict, remember why you agreed to work with your counterpart, including his success and to support him in executing those proven strategies. As a result, you will spend less time dealing with crises and more time moving towards your objectives.

Understand that your actions may have unintended effects. Even if they are intended to prevent conflict or avoid escalation, some of your efforts can have unintended consequences. Therefore, choose your strategy carefully and in great detail. Know what could go wrong and plan to address this negative event when and if it occurs. When things do go wrong, do not seek to cast blame. Instead, move in good faith towards a mutually beneficial solution.

The long view is the best approach. Conflict wears on an organisation, draining resources and morale. Taking on significant risk may be necessary in a crisis, but not at other times. Finding a way to make it work with the counterpart owner or contractor can mean that your company lives to fight another day or gets back in action faster. A contractor or owner who takes the long view can incorporate Master Tzu's lessons on finding meaning in paradox. For example, "being calm in the face of calamity" and "finding order in disorderliness" are two principles. When conflict arises, a calm and measured approach focusing on collaboration rather than competition will often make a significant difference. After all, just like the ancient Chinese society, a company that plans to be around for a while must also deal with occasional disagreement. Testing potential leaders who can stay calm and contribute to the solutions is a good strategy. Adding to calamity or disorderliness due to one emotional reaction only adds to the problem. Being objective is best. A contractor might use quantitative analysis to keep emotions out of crisis management in our business.

Our business is not as serious a matter as the military, but this book points out timeless strategic approaches regardless of the endeavour. A great company seeks not to create a wholly unique strategy but takes proven and timeless concepts and builds upon that wisdom. A construction professional might call it, "Standing on someone else's shoulders". This method is efficient in creating effective strategies and is also more resistant to short-sighted decisions. Furthermore, starting from proven concepts allows more time to tailor a strategy to the given company characteristics. The Art of War contains many proven and timeless ideas.

Built to Last: Successful Habits of Legendary Companies

Jim Collins and Jerry Porras's "Built to Last" has been one of construction professionals' most talked about books in recent years. It deserves this attention. The book has been cited as one of the best business books by The Wall Street Journal's CEO Council members. The book's value to general business is clear to many; however, we assert that the construction industry has a unique combination of factors that are not repeated in any other industry. This section analyses the applicability of this book's lessons to the business of construction contracting.

This conversation is timely. The challenge of creating an enduring and sustainable construction firm becomes more critical as the Baby Boomer generation retires. The cumulative number of those at the end of their business career grows with each passing year.

The book's purpose is compelling: to identify the habits of well-established firms that continue to outpace the financial results of competing firms significantly. Interestingly, the authors identify and analyse many successful firms from various industries, but construction contracting is not among them – even though it represents the largest private employer in Australia.

The Book's Alignments With the Construction Industry

Great management books have optimised value when they are applied. See 4Mat Teaching and Learning. The formative teaching and learning approach is a robust and efficacious process to capture learning gains.

Data Should Drive the Discussion

Sherlock Holmes once stated, *"Data! Data! Data! I can't make bricks without clay"*. *Built to Last* uses data as its building blocks, which is in contrast with other management books that tend to focus more on anecdotes and impressions. Accurate data is the best starting point for serious discussion about potential strategic or management action. Opinions are easy to form and assert.

Data Makes Fact the Driving Force

Careful use of data makes management and operational decisions less discretionary. Collins and Porras have made this the foundation of their research. If collected and tested carefully, data provides a stable foundation for decision making. Quantitative methods are used to bring disciplined thought to any journey of discovery. For our industry, accurate capture of data and quantitative approaches to analysing it has all the potential to start a significant change - one we think is overdue. This should lead to better decision making and greater professionalism.

Data Can Speed Decision-Making

Capturing accurate data and analysing it deeply allows us to determine our course of action quicker. Standard measures give us known facts. We become comfortable with their source and meaning through our use of them.

Data Removes Opinion From the Discussion

As construction firms are paid for performance, the numbers help us keep extraneous factors out of consideration. It is objective and fair. As we said, using a quantitative method is effective to manage your business better. However, please be clear about what it does. It *starts* the conversation about what is occurring and what can be done. The decision you make is yours, which is the art of business.

W. Edwards Deming said it better, "Without data, you are just another person with an opinion". Our firm's use of data to establish levels of efficiency is in the same spirit of *Built to Last*. Depending on the firm and its situation, contracting efficiency can be measured by comparing it to their market sector peers' average overhead to direct cost (OH/DC) ratio.

Culture Matters

All companies have a culture –a shared set of beliefs and practices. A company's culture is reflected in many ways, some obvious and other subtle such as the management style of the owner (benevolent dictator or collaborative leader), the interaction between the employees (competitive or compassionate) or the degree of formality (rigid reporting hierarchy or informal matrix organisation).

The book offers a dataset that ranks visionary and comparison firms in 6 core organisational characteristics: 1) "Core Ideology", 2) "Cultism", 3) "Purposeful Evolution", 4) "Management Continuity", 5) "Self-Improvement" and 6) "Big, Hairy, Audacious Goals (BHAGs)". Each of these can be categorised as a cultural characteristic. In another way, these are actions with which a company moves forward on its journey. Some are foundations to keep stability and others are prompts to improve.

Culture is also a default to which employees start their thinking. If the default is fear of recrimination, then many phone calls to those who have power (informal or formal) start the process. If the default is freedom of action based on sound reasoning, fewer phone calls and more planning with a few trusted others begin the process. Every company has a culture that affects how employees relate to each other, their leadership and their workplace challenges.

For construction contracting, we think that *Built to Last* offers some important lessons for generating and continuing a great culture:

Articulate a Core Purpose

The book suggests that a firm's core purpose is critical to the operating culture. It answers what we do that is more impactful than just the products and services we sell. For 3 M it is "solving the unsolvable problem". For Granite Rock Company, it is "making people lives better through increasing the quality of man-made structures". This vision may be to erect monuments or generate a quality and integrity legacy for construction companies. These types of statements give inspiration and direct that inspiration purposely. The company makes sure they apply in good times and bad for the company and its individuals.

Prioritise Fit Over Credentials

Inspiring and ambitious goals and visions provide employees with something to get behind and support. But, like the authors' designated great companies, superior construction firms select employees based on their character and compatibility with the core values rather than credentials.

Many efficient companies we have worked with have a "competing against yourself" culture. As both a company and individual ethic, this is a robust culture to keep a controllable action in front of the company employees. In construction, these principles work well.

Leaders Come From "Deep Inside" The Company

Leaders who are well indoctrinated in the company culture can tap into something new arrivals cannot: how the company communicates and employee trust is earned. Collins and Porras's idea, "Deep Inside Leaders", reflects a person's long history of working within acceptable boundaries of the firm. Promoting from within generates a continuity of culture and also rewards loyalty and longevity in the company.

Meaningful Change Comes From Deep Inside

When change is necessary for a firm to grow, diversify or change, the "Deep Inside Leader" is well-positioned to lead this change. This leader knows what, how and at what pace change can occur. No matter how worthy the difference, changing a firm will be less disruptive if it is directed with an insider's knowledge. For a construction firm where attention to detail and small margins mean the difference between staying in business or folding, driving change from outside is too risky.

The Book's Misalignments to Construction Contracting

All the companies listed in the book are fixed cost types of businesses. They have a business in which the product cost is less than 50% of each sale. These firms generate contribution margins from each sale up to 90%. For example, construction contracting is a variable cost business. A construction contractor's project cost is upward to 90% of the contract. As a result, the construction firm's contribution margin slides to 10% and lower. Fixed cost businesses enjoy an enormous financial cushion on each sale to do such things as stockpile talent, pay for extensive experimentation, and reduce price when demand slows.

Contribution margins of 5% mean 20 times the turnover must make up any mistake to break even on the loss. In contrast, 90% contribution margins mean that any oversight is paid for by a little more turnover than the error's cost.

Each industry studied does not perform custom work. They create a new product or service offering and then replicate it in the millions. In contrast, contractors build custom projects with some (not all) new partners in a unique place. The variability of factors (large and small) is much more. A significant percentage of them are uncontrollable.

The best run construction firms' Returns on Assets exceed 40%, just as the best companies in *Built to Last*. However, in our observation, the way they do it is significantly different from fixed-cost businesses.

Construction is a craft-centred (not machine-centred) and management intensive business (custom projects are the norm). People are the great "wild card" Construction projects can only be built well by those who have years of experience.

Unfortunately, these people are not quickly trained nor graduating from college industry-ready.

The book's featured companies can offer continuous employment and thus grow their culture deeply. Their messaging and examples can be daily to the same set of employees over many years. However, construction cannot offer continuous employment due to its nomadic, cyclical and seasonal nature. In our experience, companies' multiple layoffs hurt the loyalty and culture of the people who physically produce the work.

Many visionary and comparison firms also have the luxury of inventory and other fixed assets that can be liquidated (and, if needed, at significantly reduced prices) in slow times and increase cash flow. These fixed assets do not have to be replenished until the economy recovers.

This is not the case with construction contractors that do not have inventories of ready-made products or many valuable fixed assets waiting to be installed. Additionally, the construction industry has historically been the first that slows in economic cycles leaving less time to plan for a downturn.

Geographic challenges exist in construction. Most employees are located on the project. That project is a unique location in which work will be completed in a relatively short time. Since approximately ¾ of all construction firms are labour intensive (subcontractors) firms, they directly employ craftspeople, operators and labourers to produce finished work. Additionally, they build many projects at a time. Construction's practice is chaotic in comparison considering all the book's studied firms typically own (operate long-term) the work location, i.e., building or factory where their business operates.

A brand name is an illusion in construction. With each completed project, it grows stronger, stays the same or weakens. People are the most significant variable and input. If they are first-rate, then clients believe in the brand. However, the brand does not continue if people are not constantly supported by leadership. People are the enablers of the product and they are unpredictable as a group. Many years are needed for craft and management to gain expertise. As we have stated, humans are the great "wild card" in construction. Each is on multiple professional and personal journeys. Their focus can wane due to distraction from the one journey that matters to the client: the project outcome. A good brand name for a construction firm results from the customer's safety, quality, timeliness and final cost. *Built to Last* asserts that brand name is built over the years and has staying power through a cult-like dedication to the core principles.

A construction firm's completed projects number in the dozens in any year contrasted with visionary company's sales units are in the millions. For construction contractors, this dramatically increases each project's operational and financial significance. Therefore, each management step is critical on the path to a completed project. It requires management intensity to ensure that the one-time opportunity to gather complete information, execute detailed planning, coordinate resources, and schedule realistically is not missed. Culture helps you get there, but management enforcing discipline ensures the completion of each step.

Clients' cost is under constant review since it is a top-three business investment or the highest personal investment.

In construction, companies can stay the same sise for long periods and be profitable. One reason is that approximately 90% of the cost and 100% of turnover is earned on the project site. That is where focused support should be. Frontline people are not innovators as a group. They rely on trusted methods, some of which date back decades. They have learned through "scar tissue" to be sceptical of gee-whiz ideas.

This appears to contradict the books' emphasis on being constantly innovative. Construction's reality is that contractors build custom work in a unique place with many people they do not employ, working with them in hundreds of interactions with a 4% average net profit on a deadline. Rather than "preserve the core and stimulate progress", many construction firms should instead "preserve the core and focus on perfecting a known craft". This perfection produces a competitive cost and delighted client. In our estimation, the craft which has (and will continue to) decreased in availability is what clients want and will pay more for. Leading and managing skilled craft workers is a talent that highly efficient contractors possess.

Neither Alignment nor Misalignment: Big Hairy Audacious Goals (BHAGs)

In the book's vocabulary, visionary companies are "clock builders and not time tellers". That is, they have the technical and business understanding to produce the desired result. The results evidence itself many times in profit and repeat business. Contrastingly, the time tellers are the trackers (financial, consulting or accounting people) who capture and analyse the effect. They are not the executors but can accurately point out poor results. How to remediate these results is not in their training or job description. Only the clock-builders know how to correct.

Taking big, audacious risks is generally not advised in construction. The construction industry has enough trouble. It rewards those who perfect their craft skill and operate with excellence since this lowers the overall cost to the client.

However, some of the lessons of *Built to Last* can be an added value to construction companies. If we follow the idea of BHAGs, we might ask the question, "What directions are both profitable and visionary for construction contractors?" Some of the current ones we think have high value:

1 Creating a Women-Inclusive Environment

This kind of culture is uncommon in construction contracting firms. However, it solves the talent shortage facing all construction firms. We suggest that this is a powerful advantage in the "war on talent". Since there is a small but significant group of talented women, we envision a purposeful move to make a company culture welcoming to both genders has several advantages.

2 A Generative Safety Environment.

This is a culture in which employees follow the best safety practices and create them. Executive management is a partner in this process. The employees have taken over and mentally own safety. Each person evidences it in the firm "looking after each other". The morale, financial and competitive advantages of this kind of culture are significant.

Many others are possible; however, each BHAGs raises the firm's level of value to its client. Whether through total cost (longer-lived product, quicker delivery, fewer resources needed, lower initial cost to purchase or more sales from the client's client).

Each firm can have an exciting discussion about the possibilities, but the value is in the execution. In this last step, comparison companies seem to fail.

In summary, *Built to Last* is a solid research-based book. Its interpretations rely on accurate data collection, suitable analysis methods and knowledge. To be clear, the researchers have the body of experience and analytical training needed to evaluate possible conclusions expertly.

In general, the book suggests that each company has created a set of cultural defaults between executives, managers and front-line employees. The quality of those defaults in both depth and breadth determines performance. Visionary firms appear to have spent a long time thinking through the effect of each cultural default. Some may have been arrived at intuitively. These prominent firms have leaders of stellar character who show by example for each employee to observe. Some of these leaders are formal, while others are informal (also known as "thought leaders"). Subsequently, many of these employees have decided to invest their emotion and intellect in their work for these leaders.

We believe an employee's emotional attachment is the critical piece. It releases each person's intellect in full force. Positive emotions become good behaviours such as working long hours, attending to detail and persisting in problem-solving. Leaders set those emotional defaults in many ways. The best ones use all the methods possible; however, it starts with the mentality of leadership: "we succeed or fail together". The result is that all employees default to the company's core beliefs, no matter how tired, stressed or flushed with success.

There appear to be no dramatic moments as companies become visionary. It is a continuation of steady and unglamorous work by company members. Day-in and day-out, leadership prompts employees to think in specific ways. Keeping this in mind, the visionary companies "get out of the way" of competitors who take huge untested risks and fail. Each time they lose the confidence of clients. The consistently good and highly effective company reaps an increase in the competitor's client business.

Our overall assessment of visionary companies, as described by the authors, is that their leaders capture and retain the full attention of their employees. This connection can be described as "sticky". From that, these executives possess the full force of their staff's intellectual power. Then, each visionary company's leaders direct that power in ways that increase value for the customer. Each customer reciprocates with more orders. Sometimes, they help these leading companies see future changes which enhance their present success.

As a sequential process, the general lessons of managing a construction firm and making insightful decisions should first be governed by data and fact, then interpretation and opinion and lastly, filtered through the company's character (culture). The lack of consistency of this process causes others to fail in their overly ambitious and untested plans. Conversely, this leaves disciplined firms an opportunity. Conveying these lessons is the intent of Collins and Porras's *Built to Last.* We believe that they have succeeded.

The Tipping Point

Malcolm Gladwell's contemporary book, *The Tipping Point,* explains the roots of social and medical epidemics. Although interrelated, one is desired by all and the other by no one. For businesspeople, analysing the social epidemics of our times assists those who want client acceptance about their business value. This is important if we are to "recession-proof" our contracting company or career. The book is one for the ages. Gladwell reaches back to Paul Revere's ride and narrates forward through Bernard Goetz's vigilantism, Hush Puppies' newfound acceptance and the AIDS pandemic to the present day.

Marketing in construction comes to mind when reading this book. However, it is not traditional marketing in the general business, but the best marketing a contractor can enjoy. It is not the web, the latest "tweet" or meeting event. It is quite simply a positive word of mouth.

The construction industry is dependent on people. That has not changed in centuries. Today, clients, designers, employees, suppliers and service providers all make for solid word of mouth about your firm and yourself. Great work is appreciated by many; however, how fast it travels and how deeply it is ingrained in others is the point of the book's instruction on social epidemics. Tipping it in the right direction can make a significant difference in business.

Tipping Point adds to the body of knowledge for most business people, contractors or otherwise. It is not a mainstream operations book but a look at how different trends are started and kept alive. Indeed, a business fad has significant impacts. The subtitle is "How Little Things Can Make a Big Difference" is true. If your parents said this, they were right.

For each of us, all outcomes have causes. For example, if a person has a poor business reputation, they have earned it through an inattentive focus on effective business activities. Mediocre grades are the result of weak scholarly habits. Financial distress usually results from less saving and more spending. These kinds of events (effects) in our lives have roots in other events (causes). Be sensitive and honest about these causes and the resulting circumstances. The Easter Bunny is not the villain. Once you realise this, you will experience more good outcomes than bad over the long term.

This seminal work opens up the mystery of going from "cool" to "hot" word of mouth. Spreading it quickly is one consideration; however, the quality of word of mouth is another. Both are equally important if a professional is to enjoy an efficient journey to their desired business destination.

The book analyses things such as word of mouth and trends, go from a modest pace to a frenzied one. The application of this book for contractors seems to be for marketing, an external process. However, it applies internally to a company as well. Several questions are answered in its pages, "How do we make our new strategic vision real and quickly adapted by our people?" and "How should we position our program of change, so it is "sticky" in our employees' minds?"

To enjoy positive word of mouth in construction contracting is to have a great craft skill in building work. That is the only starting point. There are no alternatives. Clients who enjoy quality work delivered on time with no safety problems at a competitive cost tell others. In a phrase: low maintenance construction firms delight their clients.

Getting that truth spread to others quickly has great promise for the contractor's future. Many inquiries lead to many bid opportunities. With more bid opportunities, a contractor may raise their price and make a higher profit percentage at the same volume.

Longevity has many benefits that many young people may not appreciate. It affords good things to happen. But, as we know, good things take time to happen.

Malcolm Gladwell articulates this elegantly. So, do what keeps you in business: financially conservative, carefully estimate, exacting quality standards and engaging people skills. Do not forget, the longer you stay in business, the greater the chance luck will find you.

His book is one of more science than art. He breaks down exponential growth of a trend, fad, epidemic and even a business organisation. Another way to look at Gladwell's work is to understand the concept of "Aim small, miss small". In the movie, The Patriot, that maxim meant to focus on a specific target area that to miss meant only by inches and not by yards.

We are primarily interested in the latter for the construction contractor, although the former is good for dinner conversations with our loved ones.

Change happens each day in construction. Some of it is good and some of it is poor, but all of it affects your firm. So, when the good happens, take advantage of it.

The Law of the Few – people, are critical to spreading good information about your firm and you. Gladwell categorises them as Mavens, Connectors and Salespeople. These three help spread the good or bad word in different ways. So being strategic about specific types of people (Mavens, Connectors and Salespeople) makes for an efficient business "epidemic". Then targeting persons who represent those types can significantly tip the balance in your favour. Contrast this with the current thinking that "the bigger, the better" social and business network. Without a compelling reason to remember you, people will forget you. However, a central core of business acquaintances can boost your firm's profile.

Let us review the three types of persons outlined in The Law of the Few.

Mavens – these are people who sincerely want to help and typically know good information. They are forthcoming with it. Their actions give rise to trust. Mavens have no goal other than to assist. All the while, they take an interest in acquiring helpful information. They build a great social circle in time.

Paul Revere was a Maven. The colonists trusted him in great numbers. This occurred over time as Revere acted in helpful ways. To be sure, he had social gifts. When he streaked through the towns of New England, people knew him and that he should be taken at his word. Gladwell contrasts Revere's ride to William Dawes, who also rode in another direction the same night, shouting that the "British were coming". Dawes alerted only a few. Many rolled over in bed and ignored him. Those Colonists did not know him or trust him deeply. Revere had a great number of people who knew him where Dawes had much less recognition.

Connectors – These people are most important in achieving a word-of-mouth epidemic. They belong to many social groups. Connectors know more people than average and they have an affinity for building a positive relationship. They have acceptance of people who have different backgrounds and needs. Even where they have a weak connection or "tie" to a person, they can connect. Convince a connector of your worth and they will do most of the work of spreading your good message.

Research about connectors by Milgram and then later by Tjaden shows how people can be just a few relationships away from knowing each other. The "Six Degrees of Separation" is the idea that most of us are just six relationships away from any person. You know someone who knows three persons removed from the person who knows movie star Liam Hemsworth.

The connectors know so many people in diverse social groups and have earned trust by many of them that they minimise the number of relationships to get to that target person (or audience). They can spread word of mouth faster since they maintain this unique position in business circles.

In this way, Paul Revere was also a connector. He belonged to several diverse groups (intelligence and anti-British) and was tolerant of people's differences.

Salespeople. Salespeople persuade. They convince others of an idea, service or product. Sometimes with enthusiasm or just by caring, but they convince others. Said differently, salespeople overcome others' indifference or natural resistance to changing their thinking.

Most of us have experienced a salesperson in some way. As we just said, their goal is to move your thinking to their side. Depending on the approach and the quality of their thinking will determine the speed of change, if at all.

The clock is ticking in all of our careers. We want to be introduced to the great client(s) who will help our business along. Working smarter about critical players in any market is time well spent before engaging in marketing and sales work.

The Pareto Principle is alive and well in Australian society. As the great Italian economist once postulated, 20% of most populations commit 80% of most acts. In real terms, 20% of beer drinkers consume 80% of beer, or 20% of motorists cause 80% of accidents. This 20% are essential people if one is looking for the largest source of any activity. Control these folks more, directly affecting the event (such as beer consumption or car accidents). They make activities exceed the tipping point or not.

On a positive note, it is also true that 20% of the population (influential contractors, designers and owners) is instrumental in making things happen (projects funded or built). Get to know that 20%.

Gladwell introduces the *Stickiness Factor*. He defines it as making something or someone memorable. Marketing people call this many things. Differentiation is the most common. You occupy a parking space in the head of the client and they do not forget. Your impression is indelible if your stickiness is high.

The Power of Context shows the effect of condition or circumstance. As a simple example, if a well–respected person introduces you, that context immediately gives you a greater acceptance. Alternatively, you have a poor outcome with a troublesome owner; there is forgiveness by most people due to a knowledgeable industry (sympathetic) of that owner.

An interesting aside is Gladwell's explanation of the *Fundamental Attribution Error* or FAE. Simply put, people incline to blame or credit an individual for the result rather than their circumstance. Grid Iron quarterbacks come to mind. A great supporting cast and a competent play win the Grand Final in an otherwise difficult career with awful teams. A savvy estimate coupled with a flexible client and a project manager can look golden in our construction world. Reverse the circumstances and they might lose their job.

Predicting a new emerging market segment is one lesson for the construction industry. Prepare for success by forecasting the future. With recent developments in Wind Farm, Solar Energy, Fracking and Real Estate investment, those types of contractors have enjoyed growing business.

We hope other trends emerge, such as the importance of quality work (versus low price) re emerging as a priority in the construction market, more than 380,000 construction firms will be rewarded. Then, finally, they will be given respect by buyers.

House flippers are gone and the go-go days of overleveraged real estate development may not return. These people who were the masters of the universe in 2005 are not well respected now. These kinds of business epidemics come and go.

The best example of context is after a construction contracting scandal rocks a community. Construction service buyers are all of a sudden concerned about the right things: licensing, safety, ethics, financial stability, insurance, etc. It is a paradox, but true. Great contractors want others to be reminded (by a scandal) about what a great contractor looks and acts.

Size of Social Circles and Business Organisations Matter

Too big of an organisational unit is a problem in most cases. There is an upper limit to the number of people who will engage, interact well and be productive. Examples are all around us.

A non-construction industry example illustrates one lesson of the power of context. Gladwell narrates the operational approach of Bill Gore of Gore-Tex. Many construction people are familiar with this company as contractors and their employee pursues hunting as a hobby.

Gore believes in a maximum size manufacturing plant of 150 people. This is their rule. The number is significant to remember. See below.

There are accepted limits in the construction business concerning human resources. Here are three:

* CEOs should have no more than 12 people reporting to them.
* Management committees should be no larger than six people.
* Managers should have no more than seven people reporting to them.

These are not social networks but supervisory ones. All human networks have limits (see below).

However informal, these are rules of thumb, many of our clients swear to their value. The point of limits is not the number (yours may be different from mine), but there is wisdom in using them.

What is the limit to the size of site personnel? How many project managers should be housed in a department? These are human resource questions; others are ones that a strategic plan may answer.

The problem with many employees is:

* More people cause less mental ownership by those people.
* Each person feels less that what they do matters.
* More people are less manageable.
* People who know each other well work better than with strangers.
* More people will mentally and physically wear out a manager.
* People need to be managed both personally and professionally.
* More people dilute accountability and responsibility.
* Easier to blame others.

So, Gore-Tex has found the rule of 150 people to be effective. It keeps things manageable and moving forward. In one way, the messages he sent were more manageable.

He could isolate and concentrate on a specific factory and make sure that they understood. From there, he could move on to the next.

This is not an arbitrary number. The 150 is statistically accurate. Gladwell reports that this upper limit is supported by research. In the medical world, the "neocortex ratio" of Homo sapiens is calculated to be 147.8. This ratio signifies the number of people with whom a person can have a truly social relationship. That is quality relationships with. So, there is little need to have as many as possible; people are rational and know that lightly knowing someone is not a reason to utilise their service/product or to recommend them.

This has application for the marketers and business development professionals of the world. Those who choose to have a vast network of people as a business principle are mistaken. Gladwell's lesson is that a professional must choose 150 well-connected people who will be an asset to them. This is hard work. The mindlessness of extensive, blind appeals to anyone who will listen is very inefficient in the long term.

However, as we well know, those large groups are important influencing others – whatever our goal, our job is to reach those important people with our message.

Indeed, the most quality-minded construction services buyer is not the most vocal. They would be harmed by their favourite contractors being distracted by another client. Anecdotally, we have witnessed little word of mouth among these types of buyers. So, it is up to you to spread the good word about yourself.

The Three Rules for Exceeding the Tipping Point

The idea of business networking sites such as LinkedIn is solid. You will eventually link to a Maven who will carry your message to many others with each connection. If they are an advocate of your firm, they may do so quickly. The challenge is making such an impression that the maven carries it to others quickly. For example, if you have well-substantiated truths such as prominent references, certifications or completed projects, these mavens will bring them to others rapidly.

Lessons for construction contractors from this book are:

1   Have compelling and positive truths about your firm.
2   Plan how you will communicate these truths.
3   Communicate these truths to strategic individuals. These strategic individuals should include mavens, connectors and salespeople. Each has a function in your overall word of mouth campaign.

Many sophisticated marketing newsletters go out once a month. There is a compelling business reason. After two years or 27 messages about any company, people make up their minds. To send your message 27 times in 27 days is annoying, if not contemptible. Furthermore, E-newsletters sent out often tend to be classified in the "spammers" group. This can be the death of communication about any good product or service offering.

The person on the other end of an email gets a vote. They rightly can make up their mind about any company without a daily reminder from that company. Annoy them and you have lost them. They may tip the client community the other way against you.

There are two areas of work acquisition that are helped by positive word of mouth, 1) Small projects and 2) Sophisticated buyers of construction services.

Small projects

Small projects never go away. They will stay in the project mix whether the economy is healthy or not. But, as we know, Giga projects are cyclical and rare.

Most small projects are not advertised heavily, so word of mouth among those who fund and own them is essential to acquire a steady stream. Small projects do not delight most contractors, but they keep people busy and productive while their ROI is above average.

Do not discount the power of small projects. They possess three desirable characteristics:

- Higher percentage of profit.
- Faster payment cycle.
- They will never go away.

#1 and #2 produce a significant ROI. This is a crucial characteristic of financially successful firms.

Sophisticated Clients

Clients who have built projects for ten years or more have a discerning eye for the quality of construction services. This is especially true for those buyers who will own and operate the projects you build – ditto for their family residence.

Sophisticated clients are the target of most tenured contractors. They will buy quality and believe that *"the sweet taste of low price disappears quickly with the bitterness of poor quality and service"*. In their experience, they have argued with the lowest bidders who delivered low prices and are now unconcerned about quality and service. These sophisticated clients are careful not to repeat the mistake. Their discernment is as deep as a well-respected construction firm's savvy is. Both parties are looking for each other.

A common term "leverage point" is what describes many parts of this book. A leverage point might be a person or a thing. A well-respected contractor is one example. "Harry said" can be all the endorsement you need. A beloved association president is another. These people can get you to where you want to go. Be a judge of who these people are in your circle of business associates. They are a leverage point and an efficient one. As they say in NASCAR, "getting gone" to the finish line first is the goal.

When we continue adding to our body of knowledge, eventually "tip" our ability to reach our goals. The business environment will make more sense. Several senior contractors have told us stories of their struggles in the early years. They barely made payrolls while experiencing call-backs and difficulty winning new work. Then, somewhere around ten years of wrestling with the business of construction contracting, it became easier. Things started making sense. They acquired "eyes that see" the company more deeply and simply.

They describe "beating their heads against a wall" to no success. Finally, they understood what they needed to "unlearn", and the subsequent profit/loss statements proved it. It became clearer how construction contracting operates as a business and a lifestyle. They endorsed it unemotionally, knowing it worked. They had the scar tissue to remind them.

So, if you are having difficulties in this business, do not give up. It becomes easier over time; it will eventually make sense if you love this business. As a friend said, "if you stay in business long enough, luck will find you".

You cannot reach the double-digit net profit or 40%+ return on net worth without a serious effort. The industry is too harsh for those who are insincere. Simply put, you will not come back with a redoubled effort after a setback unless you are committed to estimating carefully, building work well and collecting what is due to you.

Acquiring profitable work is the result of many inputs. A critical one is communicating your facts to the interested parties. If the facts are compelling, then others will signal interest. Do not overlook three influential factors 1) staying small, 2) perfecting your business and 3) communicating clearly to others. Otherwise, a company may have to "re-invent itself", often after mediocre results while on its growth direction.

So, for those who want to grow their business quickly, *Tipping Point* helps clarify the process of producing strong word of mouth with less wear and tear as we continue the full-time job of constructing work while delivering excellent results.

Thriving on Chaos: Handbook for a Management Revolution

Change is constant in the construction industry. Some may say "chaos" is a better descriptor. However, everyone can agree that past business conditions (yesterday, last year, last decade) are not the present and will not be the future. We cannot solely rely on history or experience. Whatever the degree of change in our industry demands that we be aware of it and, more importantly, act to address it. If not, the harsh conditions under which we work will make unobservant companies extinct.

Tom Peters's legendary book, "Thriving on Chaos", is an appropriate starting point. What "Who Moved My Cheese" lacked in the prescription of managing change, this book makes up for instruction. This could be a post-graduate course that adds to the "Cheese" instruction.

We guess Peters's work launched thousands of consulting careers worldwide. He led others in thinking and addressing lagging innovation in a dynamic world. He identified a quicker cycle of change in the world than previous decades and produced an unblinking book addressing it. "Thriving on Chaos" is a substantial addition to his body of work.

There is little choice in our intertwined world today but to welcome change. Executives must choose a direction of what to do in the future. Backlogs are no longer sure. Owners are neither stable in their business nor predictable in their demands. Nervousness exhibits itself in many ways in the construction market.

The good news about change is that it rewards established firms. Those who have a body of work are more trusted by clients and funders than those who are new. A substantial work history, no matter a company's resume, predicts future outcomes with high certainty. Sophisticated and wealthy construction clients are sensitive to this.

Evaluating "Thriving on Chaos" for a Construction Contracting Firm

The book speaks adroitly to a publicly held and fixed cost business. What it has is not one for a craft centred organisation such as construction. This is not a criticism but something of a misalignment. Be aware as you look for a strategic direction or business process in its pages.

This book is centred on a volume sensitive business model. More volume creates more profit. Construction is not this type. It is a variable cost and one based on the right project opportunities, not all project opportunities.

As a quick quantitative measure of the organisational focus, the example firms featured are telecommunication, manufacturing, retail, grocery, health care and airlines. They are characterised by most costs residing in the firm's fixed assets. Airlines have airplanes and gates' rights. Steel manufacturers have plants and inventory. Telecommunication firms have the technical/electronic infrastructure. To increase profits, they must rely on a few spectacularly profitable innovations. This is a good thing and a good strategy.

As a practical definition, a fixed cost business is one where the fixed cost of the business is greater than 50%. So, with each sale, the cost of executing is a minority (less than 50%) of the total cost.

The asset is most of the cost of the transaction. Each month starts with a large "nut" to pay for. Generating turnover dollars is critical for this type of business model. If a fixed cost business has paid off its liabilities (which fund the acquisition of the assets), then it can approach a 100% gross margin of its sales.

Construction is different. 80% to 90% of the costs of a firm are the costs of projects. Fixed costs are a minor portion of their cost structure. A single purpose generates all profit: completing projects on time, within budget, safely and adhering to plans and specifications. There are no other options to generate a profit. The projects are the "main thing". It is important to note that most projects are custom and not repetitive. It is a challenge that most other industries do not face.

In a fixed cost business, gross margins can be significant on each sale, so there is a great incentive to develop an innovation that increases sales. One innovation or new product can be a "windfall". Also, a company can bury mistakes under a pile of profit from a few breakthroughs. Peters's tremendously in-depth analysis serves a fixed cost business well. Sometimes gross margins on sales can reach over 90% so that innovation can reap great rewards. The energy spent on finding a better way or product is justified.

Take the example of a new use of an asset such as an airplane. One benefit is for passengers but might also be 2) for package shipments or 3) Use as a charter for private corporations. So, the idea is to adopt number two or three. The asset cost stays the same, but the turnover steam can be increased.

For example, if passengers and package shipments are sent simultaneously, turnover dollars increase and the extra operating cost rises minimally. More profit is a result.

Uncovering a way to make an asset more productive, especially a machine, intellectual property or even a person, can be a boon to a fixed cost business.

In another way, the fixed assets are visible to competitors and others. This makes these kinds of companies a sitting duck. Where private or public, a steel plant or a retail store is a stationary thing viewable upon a public visit. It is hard to be invisible and discrete. So, innovating and becoming a moving target is a necessary strategy. That is if a firm is to beat the competition and thus improve profits quarterly.

Take the firms and leaders who have appeared on the cover of Business Week, Forbes, Fortune and other leading periodicals or featured in "In Search of

Excellence", "From Good to Great and similar books. Many are bankrupt or have a diminished presence in their markets. This is not a reflection on the books or their authors; it is a reflection on the nature of capitalism. Legendary leaders live and die. Many times, their company has no equal replacement to them. Also, business conditions change, so the company's fortunes flag. "Creative destruction" is a natural phenomenon.

One way to look at Mr Peters's book is a stellar attempt to capture expert knowledge of legendary leaders. Non-business organisations such as the Government do this as a matter of routine now. All commercial industries are trying to capture their savvy manager's methods and thoughts as a significant percentage of Australians retire. This book is an early pioneering and visionary effort in this regard.

Reviewing the table of contents tells of both sizzle and substance. Six areas are covered in this book:

1   Prescriptions for a World Turned Upside Down.
2   Creating Total Customer Responsiveness.
3   Pursuing Fast Paced Innovation.
4   Achieving Flexibility by Empowering People.
5   Learning to Love Change: A New View of Leadership at All Levels.
6   Building Systems for a World Turned Upside Down.

It is clear that any latest fad captures many managers' imagination and interest. There is a market for new thoughts. Many books are sold based on sizzle and not substance. However, some are substantive such as this one.

Books cannot be prescriptive. The author cannot tell you what to do because he does not know you and is not writing just to you. So, with any great innovation, more work is needed to refine it. A serious leader must invest many hours if innovation is to be applied to a specific situation. If not, false starts will example, the disastrous lean manufacturing methodology application at General Motors. Even with many hours of tailoring, successful implementation is never sure.

Construction contracting business research is the missing piece and, thus, a potential leverage point for our industry. Since we are a primary contributor to the Australian GDP, it is unfathomable that we do not have more fact-based analysis about construction business practices. Rules of thumb and informal business practices are still standard. As construction firms stay in business, it helps our country in many ways. Remember what Pete Wilson stated, "A job is the greatest social program ever invented". As a further thought, unemployment may be a great social ill.

Application to a Variable Cost Business

By our definition and others, a variable cost business starts with the premise that there is less than 50% fixed cost in the company. So, for example, construction contracting is a variable cost business.

Most other service businesses are variable cost. Most doctors have a medical practice that is less than 50% fixed cost. They can rent space, lease machines and have flexibility about their cost structure. The number of transactions can be predicted weeks in advance from their appointment book. Construction does not have that predictable project or turnover stream.

A service business can be copied freely in most cases. This is a competitive problem. There are several other traps as well. However, there are some positives. Human enabled processes characterise a service business. Great people are a consistent asset in all service businesses. So, a core group of independently working and trustworthy employees is the answer.

However, Peters is correct in stating for all businesses, "The new market realities demand flexibility and speed". He is pointing asserts that companies must fight the NIH syndrome, i.e., "Not Invented Here". Any idea or tactic borrowed can be yours for the price of observing it and understanding it.

Applying the Book to Construction Contracting

This is a "How to" book on transformation and change management. There are dozens of ideas and processes. Select the areas you want to innovate and start your research with this book.

The variable cost and service nature of our business is unforgiving. Variable cost means we cannot have a system or product that, once formalised, dramatically increases the volume of construction. But, as we have said, we are human-enabled and people are not consistent or predictable without extensive training and monitoring.

Clients and suppliers have the same characteristics but, we cannot train them or control them. They are also critical to our success but can be a wild card in a quest for profit and organisational value.

Mr Peters correctly points out several management misalignments regardless of the company type. Identifying them is the hallmark of an observant professional. Addressing them with explicit instruction is the mark of an insightful one.

1   *The disconnect between the base functions of a corporation and its staff functions.* Mediocre manufacturers pay more to market and finance employees than to their plant managers. Those who create the product are not "voted for" with dollars as much as others performing support functions. Do we have the same misalignment for construction firms? The craft hour in the highest value in construction. Are site supervisors and project managers who build the company product paid less than others who support it? Is site experience glorified? Does the office fully fund the project operations? Again, the site operations are where most construction profits are realised.

2   *Becoming obsessed by listening.* Gathering good information and insights from others helps companies keep mistakes low and direction effective. Whether public or private organisations, it makes no difference. Engaging others cannot be dismissed as just an option. It is a critical skill any aspiring construction executive should have. Said another way, acquiring "perfect information" must be the goal of any serious manager.

3   *Do not get stuck in the middle.* Michael Porter's Return on Investment and Market Share shows the strength of being in a niche. ROI is higher than when a firm is differentiated. These rare things are rewarded, whether it is a craft skill, management expertise or occupying a highly regulated part of the construction industry.

4   *Becoming obsessed with competitors.* Peters rightly suggests that we should compete against the best. He notes it will make you frustrated in the short term

but better in the long term. As we have said, competing against a mediocre competitor makes mediocrity a habit. Those less ambitious will have an unrewarding business career. Alternatively, competing with those trying to build serious wealth allows you to charge more for your services while challenging your firm to seek perfection in its offering. Those who are exceptionally good only expect to be better tomorrow.

5   *Use of self-managing teams.* Small and independent site crews and groups led by managers who care about outcomes large and small gives executives more time for substantial leadership. They crisis manage less. Smaller teams are more productive than large ones since they cannot hide anyone. However, they are "messier" since they cannot do all things. Sharing and cooperation between small crews are necessary so the executive can drive the safety, craftsmanship, schedule and cost issues aggressively.

Weaknesses Regarding Construction Contracting

Peters starts the preface of the book with the statement, "Few would take exception that our sales forces are not sufficiently cherished". Most contractors do not have sales forces. They rely on their body of work, including reputation and financial strength. All is needed is to "communicate the truth to the interested parties". Customers pay attention to facts such as your past work or the quality of the proposal.

It is fair to say that the book has fewer insights into the construction business. We should expect this since the author's background is not in our industry. Several passages in the book discuss change, chaos and that only innovators win. The stock market and its market makers need something to talk about. A stagnant stock market (not up or down) makes no one money. So, excitement, controversy or scandal is the breakfast of the equity markets.

This book is well organised. Peters places an order to his thinking. We outlined it above. It is a good organisational sort. However, some of his assertions are not true in construction.

Here are a few samples:

1   "Quality must be judged as the customer perceives it.
2   "Do something, anything now".
3   "Spend time lavishly on recruiting".
4   "Provide an employment guarantee".
5   "Involve everyone in everything".

Although insightful for an innovatively driven business, we assert that these would slow the construction business cycle and cause unnecessary costs in our single-digit net profit business.

We have an industry that is reliant on a few specific outcomes:

1   Safety – increasing liability for clients for any worker or bystander hurt. Also, this affects cost significantly as worker compensation rates can add or subtract 10% or more for a contractor's breakeven cost.
2   Speed – clients want the shelter or infrastructure sooner rather than later.

3   Quality – specifications, plans and building codes require adherence and pave the way for payment by clients. Clients' perception is essential but does not preclude payment.
4   Cost – being frugal about costs and insistent on the higher utilisation of fixed assets keeps the cost of a construction business the least of its competitors.

In summary, all effective leaders are experts at stimulating conversation. This is one skill of many. Keeping others looking forward rather than looking to the past is one critical skill of leadership. In another way, acting thoughtfully but decisively energises a firm. It all starts with having an alternative idea. Always having a plan "B" inspires confidence in subordinates. A great idea typically is new and fresh. It prompts others to think of an "alternative" reality. Tom Peters furnishes that.

His book embodies that and, as necessary, contains energy in its pages. Look past its publish date; it covers critical areas. Do not be surprised if a few presenters and management advisors have crafted Peters's ideas and assertions into a business, even today. In other words, Tom is the father of many consulting practices, unbeknownst to him.

The business firms featured in the book are well worn or non-existent. Do not be distracted by the names. Note: they are all fixed cost businesses, most publicly held and many international. They thrive on innovation to improve their bottom lines and stock price. That has not changed since the 1980s.

If his book is to be followed, one should focus on taking Mr Peters's best ideas and reformulating them to your construction firm's eccentricities. There are a dozen or more takeaways Peters's book gives hope and instruction to those overwhelmed by change and possible strategic directions to consider. There is an answer and Tom may have it.

Take comfort in the fact that construction is one of the last industries to adopt significant change. This vice is also a virtue. The business is straightforward however, not easy. The industry inclines to trust the past (practices, methods and materials). So, having a robust innovation strategy is not rewarded by clients nor matched by competitors. BIM, Lean, Sustainability and Public–Private Partnerships are examples. This slow adoption process has been a characteristic for centuries. The construction industry treats missteps and misalignments harshly. Careful innovation is rational in our highly risky industry.

9 Case Studies

These case studies are to transfer knowledge indirectly. Billet and Loosemore (2004) found that case-based content can trigger different perspectives and integrate workplace practices with theory-based knowledge. There are several insights the reader will discover that the writer did not realise. That is the beauty of the case study. People bring their perspective as they review and analyse the dynamics of a situation, extending and deepening the lessons.

The following case studies are fictitious. Each is an amalgam of two or more experiences of ours. We have included questions and exercises in each to structure further thinking.

Case 1 – Pigott Builders

Dorothy Pigott works for a custom home builder. By definition, their firm builds a different house each time. As a result, some say they are a "luxury home builder". She replies, "Only if the economy is good". She means that their firm can take a more considerable risk is there a strong demand for high-end houses. Also, clients tend to add exotic amenities when they had good cash flow from a well-paying job and investments.

Dorothy is fortunate to be from a vacation area. She is born and reared locally. Many long-time connections here make navigating the business challenges much easier than outsiders trying to start a business in her area. The real estate here is beachfront with an intercostal river just a mile west of the coastline. There is much waterfront property and building lots are still plentiful.

Recently, Dorothy has seen some of the turnover and profit numbers from her boss. It was a surprise how profitable the firm was. The numbers were in the hundreds of thousands of dollars of net profit.

It made her rethink the idea of going into business herself. She had considered it some time ago and felt it was too great a financial challenge with little payoff. Her boss always seemed to emphasise company problems and industry crises. Now that has changed. She has seen some exciting profit numbers from her employer. The payoff now seems worth it.

So, at the beginning of next year, she will give notice to her employer. That gives her six months to get things organised so she can have a running start. Since she has been selling and building custom homes for over ten years, she felt confident she could land the business and manage it profitably to the satisfaction of her clients. Once word got around, she should have a decent pipeline of leads, if not sales.

DOI: 10.1201/9781003290643-9

The only doubt in her mind was the accounting/financial management area. It was a weakness and she freely admitted it to herself. So, she started to make lists and process maps of a proper accounting system to get a better idea of how accounting and finances should be handled. She knows that it would be a "blur" once the business starts up and becomes a struggle to fix any significant oversights. Additionally, any accounting issue involving payroll, payments, lien releases or financial statements would affect the bank, insurance company or the government. All three of which have tremendous power to upset their business and cloud her future.

A friend told Dorothy that "one doesn't get a clean trip to retirement". The friend further explained that somewhere, there would be a problematic owner, a suspicious governmental authority or a nosy banker who will ask for documentation about something. Keep things well documented. That is the only defence against something that will inevitably happen and maybe more than once a year.

So, Dorothy set about to make sure she had a good process to keep all things smooth and complete. It seemed apparent to her to have a sound filing system, a couple of large filing cabinets and an archival system. But, again, she knew when things started, there would be the pressure of paying business expenses, including subcontractors' pay requests and employee payroll. There would be little time to change the process and add capacity if she needed it. Better to do it on the front end. She did not want to get caught in that activity trap and not keep things organised.

In Dorothy's experience, sophisticated and typically wealthy clients appreciate well-organised contractors. This could be a selling point. These clients had a nose for construction firms who were "seat of the pants" types. They seem to be attracted to a more professional and efficient firm that can quickly produce needed documentation for construction loans, zoning approvals or tax purposes.

Dorothy thought about the accounts payable (AP) process. She needed a good person whom she knows personally. This would be a good idea as the person would be handling large sums and had to be trusted with that as Dorothy was selling work and building houses. She had a couple of people in mind who were local. The company's market area had a large population of folks who had lived there less than five years and several times Dorothy has heard stories of people coming in for a job and then leaving sometime later with problems in their wake. She wanted to make sure that would not be the case in their firm. She had to know much about the person at a minimum – personally and professionally.

Her AP process was based on a simple idea. If the invoice matched the delivery ticket and the purchase order in all aspects (such as count, unit price, date and signature), the bill would be paid up to $5,000. All accounts payables invoices that were over that amount would be personally reviewed by her. Typically, a scan would be sent to her email – invoice, delivery ticket and purchase order – and then Dorothy could approve it and send it back without the paper copies leaving the office. She, of course, would spot check all paperwork and challenge people in face-to-face meetings to make sure that things were tight and well. However, day-in and day-out a monitorable process was needed.

Of course, the easiest thing to do in construction is to pay earlier than the competition. This made people pay attention to her closely since suppliers, service providers and others are cash flow sensitive. Dorothy knew that if she paid bills a little more promptly, she could demand (and expect to receive) several things such as prompt subcontractor bids, supplier quotes, returned material credit, staffing of projects and the like. Indeed, any early pay discount like a 2% discount if the invoice is paid in 10 days ("2/10 net 30"). If calculated on an annual return on an investment basis, it is a 35+% loan. Dorothy knew she could use an early payment reputation as a strategic weapon, especially in a slow economy. Subcontractors would consider her a "low maintenance" builder if she paid timely. If they did the work well with her, they could expect payment.

She wanted the accounts receivable (A/R) process to be easy to manage. Dorothy knew she would have to be more involved with it than A/P. The collection of billing was the life's blood of the firm and the dollar amounts could scale up to hundreds of thousands of dollars in a single invoice. Variations would happen and had to be recognised, documented, billed and then, collected. The accounts receivable process started at the proposal stage, as she always made a schedule of values as part of it. If the client began to negotiate price hard, Dorothy would make sure the payment schedule was agreed to as cash-flow positive at the beginning of the job. This removed some of the adverse effects of any price cut she might have to agree to.

She started to formulate and refine these ideas. Her next thought was to choose outside partners carefully. These would-be people she could trust and who could be a leverage point for their business. These are vital professionals such as a CPA, a banker and an insurer. They would be necessary for their services and would be a source of information for things that work and other people to trust.

She wants to seek a supplier relationship but not just yet. Winning the first client house would be sometime after she left her current employer. Also, suppliers could open an account quickly with a maximum credit line of, say, $10,000. Since she is a known quantity in the market and personally owned a good credit score, there was no reason to believe that she would be denied an account. Additionally, supplier salespeople tended to trade on this kind of information with others (gossip may be a better word) she did not want the plans to get back to her boss.

Ethically, she struggled with the idea of what information or things she could carry with her to the new entity. Client leads and relationships were problematic issues. Her father had told her not to take any tangible (paper) with her. Her cell phone and computer listings were valuable but, were they the property of the previous employer? However, she felt what was in her head – relationships and the knowledge of others – was hers.

If clients sought her as a resource, that was their decision. Her father would agree. Dorothy suspected that there might be some crossover of clients choosing to work with her. These would probably be fresh leads and those looking for a competitive price from a less established (hungrier firm). Her dad (a retired custom home builder) shared that over her career, he went from less sophisticated (not wealthy) clients to ones who were highly sophisticated (and many times wealthy). As he grew his craft and business skill, he attracted clients who sought and appreciated those skills. As a result, the profit he could charge increased.

This was the ultimate strategic goal of any construction firm. People of wealth always had money, even in a recession. Also, they trusted low maintenance, high-value contractors and did not look past the proven ones for a low-cost construction firm.

Questions

Why was Dorothy so focused on her new firm's financial management as much as the sales and building process?

Determine whether ethical or unethical the possible actions that Dorothy could take in establishing her new business.

In what ways do sophisticated clients act. How do they differ from ones who may see price as influencing the majority of their buying decision?

Exercises

How would Dorothy go about finding new customers outside that she has not met yet? She feels she cannot contact any potential buyers she was introduced to through her previous employment. Create a process map of discovery, marketing, selling and follow-up client process. Insert in the map potential sources of information.

When Dorothy goes into business for herself, she may have an unethically minded employee. So, create a five-point ethical pledge that a new employee should sign.

Case 2 – Matters Building and Construction, Inc.

Gary Matters owns a light commercial and custom residential construction firm. He is in their fifth year of business. Gary feels good about prospects as he has survived five years in a sluggish economy. He has been modest in things he has built and turnover has reflected less than $1 million. Gary utilised subcontractors or speciality contractors to build.

Now it is time for growth. A developer he met at an association meeting has asked him if he was interested in building four mixed development complexes on the city's outskirts. Of course, he was. The project would start shortly. He could not believe his luck but felt he earned it since he managed through the five-year mark in a stagnant economy.

His wife, Aja was the "dark cloud" in all this. She was happy for him but did not get overly excited. She quietly started asking questions such as how he met the developer, whether the project would start shortly and whether Gary was the first contractor he talked to. Gary kept his cool. He knew his wife was always supportive but objective. Additionally, she quietly protected Gary in other ways.

Gary let himself get excited in any event. Finally, finally, he had something to talk about with clients. It had a promotional value and people seemed to take him more seriously.

Also, he felt that he could now fund some assets such as earning a NABERS rating, purchasing estimating software and building his own office. This could be a hallmark year. Aja commented, "It is a long upstream swim in a winding river. Keep your focus on the next turn in it. You cannot see the end. You never know where the rapids are". Gary knew what she meant – be careful, conservative and steady. That is, focus on risks and opportunities you can control.

After Gary delivered his proposal, the developer started playing dumb. Gary knew he was not a dumb man. His dad once said, "If someone is dumb, it is not playing dumb. However, some smart people play dumb. Know the difference".

Gone was the congeniality and easy manner. He called only if he wanted a clarification or an alternate price for a possible design change. He was on and off the phone within minutes and the developer never furnished any documentation. Gary would price it in writing but, he started to feel uneasy about this project. He felts a tug of war coming – the developer sometimes said, "*I thought you meant...*" or "*I assumed you included...*".

The developer stated from the outset that the four buildings would be constructed under one contract. However, Gary's banker suggested that that may not mean continuous construction. Real estate is a cash flow-sensitive business. This could tie up general conditions costs for more months than planned while sales stuttered or stalled. Also, if your firm's job site signs were on the job while there was an extended period of inactivity, people, including potential customers, would assume the worst about Gary's firm.

Gary felt he was playing defence with the developer. Gary felt he needed to play a little offence. However, he understood that this was dangerous since the over-population of contractors in the area. Gary may turn off the developer in a confrontational manner; then, other construction firms would be contacted. Maybe, the developer had already, he thought.

Developers sometimes target undercapitalised contractors. This gives them more control when problems arose. For example, if there were a design error or soil problem, the contractor would go out of their way to fix it since payroll was due each Friday and he did not have room in their bank account to bridge a missed pay request. Sometimes, the developer would suggest that if construction stopped, he could not make the payment request timely because the bank would be cautious. So, motivation was consistently high with a lightly financed construction firm to troubleshoot any job site crisis.

So, Gary wanted to get some facts, perform some calculations and determine his minimum or "walk away" scenario.

He knew the following:

1   Each building costs $2.3 million roughly with an eight-month construction schedule.
2   Gary's general and administrative rate is $14,000 per month.
3   His amenities cost for the job is $20,000 a month whether they built one building or four. The site supervisor's salary would probably not be reimbursable if construction ceased for more than a month.
4   The developer was not going to use a Master Builders Association contract. Instead, he would use an in-house "standard" contract.
5   At every turn of the process, it was "hurry up and wait". Finally, Gary would fulfil the request quickly as the developer suggested he was ready to go except for a couple of details, then no communication for up to two weeks. Still, no building had commenced.
6   The developer wanted to know much about Gary's banking, supplier and subcontractor's financial relationships, including credit line, terms and primary contact. He knew the developer would ask for value engineering ideas. He would probably ask him to review amenities and suggest ways to reduce costs.

His lawyer warned him that some developers try to get the cost as inexpensive as possible while making him highly liable with specific contract language. If anything, they would not sue him. Still, they would take money out of payments as back charges, especially late in the project when they were assured that the quality was built in and the project was on time. =complicated. It has happened to several contractors over the years.

These back charges would not be costed by item and might be inflated. It sounded like a nightmare, especially at the end of the project when there would be many reasons or problems that the developer might make an issue of something standard such as expected cracks in concrete or smudged finishes. He might say, "Look at this. I wonder what else is wrong?"

Aja told him that some people will say or do anything for money. Gary hoped this person was not that type. He did not know. The developer did not have any history. This was his first major project. Previously, he was a residential real estate broker. A very thin resume with an internet historical record. He suspected he had investors who were the experienced developers that made most of the decisions.

Questions

1   What is the optimum way to meet a construction client? What introductions to a developer would signal their value and seriousness?
2   What are the warning signs that this may be a negative situation Gary should not be involved in?
3   What is your minimum project ROI? Justify your answer.

Exercises

1   Create a spreadsheet that would determine a walkway number (your minimum ROI)?
2   What are five questions that you would want answered in writing before you signed a contract with this developer?

Case 3 – Smothers Electrical

Jerry Smothers has had enough of the delays in payments from two contractor clients. He is fed up. Their excuses are getting stale and his bills are due. Jerry has already gone far enough into his personal wealth to bridge the cost due to the late payments.

He learned some time ago to have a good network of acquaintances that could help him with timely information. Some of the people were his friends and some were just folks he would call from time to time to keep in touch with the latest news about what was happening in the construction industry. It was a two-way street; he tried to be helpful with information that could help them. This kept the relationship alive and active.

He knows that the two main contractors have already been paid on the jobs he subcontracted with them. Jerry has heard this from a person in the owner's office and another person at the bank. Not anything specific other than the progress payment was processed and electronically sent 14 business days ago. Jerry would never ask about specifics. It would not help him collect. The client would just come up with another excuse, then accuse the bank and the company owner of breaking confidence. As a result, people might lose their jobs.

What Jerry could not figure out is that these two customers have usually paid as all others paid. Now they were holding money for a longer time. Retention was getting held longer by a few other main contractors even though the buildings were occupied for six months and the punch list was complete on Smothers' work.

Jerry talked to his attorney and asked about prompt pay laws and other legal issues that he could force if he had to. His solicitor said it was too early to pursue such an action. However, he suggested that if Smothers was not paid after three months, the lawyer could write a letter and start building a case for a suit.

Jerry understood from his legal counsel that the practice payment delay is getting more common. Several main contractors have said in various ways to his solicitor that the interest on payments from a year's worth of jobs can produce enough profit for a typical project for that main contractor. Jerry sees red. He worked hard to complete the project with safety and quality and now he is not being paid because the firm executives see this tactic as a good business strategy. He thought the bonus system at these two firms must be predicated on year-end net profit, no matter how it was achieved.

Jerry has subcontractors to their firm for low voltage. They were paid already because that was the understanding. The suppliers were also paid in full. Some of them gave early payment discounts of 2/10 net 30 and the switchgear manufacturer gave a 5% discount for paying upon delivery. This made financial sense to keep his costs low and, therefore, bid competitively.

Jerry considered placing the suppliers and subcontractors on the same contractual condition on each project. Whatever the clients' contract terms, then the suppliers and subcontractors would be under the same. This would be cash flow wise but, would it be profitable? If Return on Investment is a critical metric, then what would be the effect? The amount of the investment determines ROI, the amount of the return and the timing of those two. Several sub-factors included payroll investment, retention, rented equipment, owned equipment and material costs. However, the equation is relatively simple in the general sense.

Both main contractors hired full-time Chief Financial Officers within the Past 18 months. From that point, things got less friendly with each firm. They said they were instituting more discipline in the fiscal operation of the business. Checks were never cut early when Jerry needed help. Retention has been held for over a year on several jobs. It seems that there was significant money sitting in the hands of the main contractors. Was this a risk aversion strategy? Somehow was there an investment angle in all this? Were they making "money on money"?

Jerry knew it cost him in several ways. So he opened a spreadsheet and started to list the costs: (Table 9.1).

Table 9.1 Retention Cost of Matters Building and Construction

Date	Borrowing from bank	Interest Payable on Balance	Lost investment opportunity – (ROI on Current Operation – 22% + interest paid to bank 7%) = 2.42% per month
3-1-20xx	$40,000	$232	$968
4-1-20xx	$60,000	$348	$1,452
5-1-20xx	$50,000	$870	$3,630
6-1-20xx	$40,000	$1,102	$4,598
7-1-20xx	$20,000	$1,218	$5,082
8-1-20xx	$20,000	$1,334	$5,566
9-1-20xx	$20,000	$1,430	$6,050

Given that profit was 4% for the firm, to pay the interest of $1,450 took $47,850 of revenue. That meant much extra work in both site and office operation just to break even. Jerry wondered if he were going backwards with poor personal credit, lack of trust from industry insiders and possibly losing some of his personal assets if things went wrong. He had signed a personal guarantee with his bank and material suppliers, as most contractors have done.

Questions

1   Is late payment, including retained amounts holding a common occurrence in today's industry? Why?
2   What are some strategies to receive timely payments, including retention release?
3   Do you predict that this payment trend will get worse or not in your career? Why or why not?

Exercises

1   Create a process map for billing and collecting payment, including physically meeting with the client.
2   Build a spreadsheet and calculate all costs caused by delayed payment. It should contain working equations such that it shows the dynamics of payment timing.

Case 4 – Havener and Sons Concrete

Joey Havener has two sons and all three own a concrete construction firm. Joey, the dad, owns the majority and his sons less than 10% each. The company is doing fine; however, the economy has slowed and things seem to be changing – there might be a new way of doing business. Joey is trying to find what is going on and what to do about it.

Cash seems to be in always short supply. He never can get a handle on it. Sometimes he is in deficit and needs to borrow from the bank and other times he taps into his personal credit card and personal savings. He knows it costs him in two ways. 1) In interest and fees. Banks seem always to find a way to insert a fee on top of interest costs. 2) His sons' lack of appreciation for fiscal conservatism ("money does not grow on trees"). They lack the understanding of the stressful early days of their business when they were children. Joey always comes up with the cash when needed. He has a distinct impression that they think this business is highly profitable and do not need frugality. He vows to start a change campaign to fix that thinking in his sons. He was bothered by one son's comment that used contract turnover and profit interchangeably. The son seemed to think that these terms were the same.

Joey thought back to how he learned from his dad. The answer was, there was not a formal learning process. His dad just made him "do" and not think. One time, his father pointed to a horse and said, "let him think for you". Joey did not like this but was fearful of his father that he just laughed and went off to the next task. The learning process mainly was osmosis with quick lessons from an impatient father.

His relationship with his kids is different. They have two-way conversations and not one-way lectures. That was good. However, they do know that have accounting and finance is critical. The company has someone who does that. So, they have not struggled mightily, almost losing the company in the early day. Those struggles made Joey promise himself that he would be financially careful if he ever had money.

He felt that setting goals with his adult children might be the answer. Although placed on an incentive plan akin to the finance industry, i.e., small salary and significant incentive for success, the boys would get the message after a couple of months of financial stress. They would see a small number in their checking account with a weekend coming up. A few weeks of that and they would be hyper-vigilant about costs and collecting billings.

Distributor salespeople were friendly; however, they seemed to want to talk to the boys more and Joey less. He never took a factory trip to one of the vendors. However, the boys have been mentioning this lately to get to know products better. Concrete supply trips always included hunting or fishing as part of the schedule.

Joey knew that he would have to be stern and thoughtful to make sure his sons did not lose the gift of a profitable construction firm. Part of his thinking was that each son loved the business with different intensity. He thinks that the one son with a less serious attitude might need to do something else. Indeed, a friendly buyout for shares but, then the company would have a less variable leadership. A second son might have to decide whether to continue working in the firm. That might be a decision made due to a renewed focus on primary duties and lack of adherence to those duties; boring as those tasks may be, close adherence to them is a good test of a person's passion for the company and the industry.

Questions

1 What are observable behaviours that a person exhibits when they have a passion for a job and a company?
2 Is Joey just being paranoid about his sons, or is there compelling reasons for him to be concerned? Justify either answer.
3 What questions should Joey ask – positive and negative –lead his sons to the correct answer – stay or leave – for each of them?

Exercises

1 Joey is very aware that his days managing the company are limited and it should have a continuity plan. What five processes would increase the probability of long continuity? For example, a board of directors that are non-family members who could oversee and decide significant issues? Write a 5-bullet point memo explaining each possible area.
2 When it comes to transferring ownership of the firm to his sons, how might this be best accomplished? Would a sequence set of exchanges of money for shares over the years be best? What might be an appropriate period? What are the pros and cons of this process?

Case 5 – Dann Electrical Contractors

Michael Steele is a middle-aged chief executive of an electrical contracting firm. He is growing their business with the help of a majority stockholder, Persevere Capital Management. They recently purchased the firm and have brought in

Michael to make the business an elite one in strategy and operations and as a result, increase its profit and organisational value. Michael is unusual in his background and credentials as a contracting executive. He holds a PhD in construction management and has been in the business since his teens. He loves the business, its people and its nature. He feels it is the best industry of all. It is obvious to him in many ways. For example, 1) craft skill is rewarded 2) competitors make basic mistakes leading to a greater competitive edge 3) High volume is not needed to be highly profitable 4) The business will never be exported overseas. There are several more.

As the new Managing Director of Dann, he has slowly understood the firm's unique characteristics due mainly to the people, past and present, whose decisions and culture have become policy and procedure. He is now trying to unwind some of those policies and guidelines without causing too much disruption.

His area of focus is the accounting and financial management functions. There are several industry-standard methods and others unique to Dann that Michael wants to change. He has done much work in "best practices". He believes that the first rule of best practices is that there are few in construction. Most people, including management consultants, use the term loosely and never have done data-driven research to correlate them. The term "best practices" sells seminars, consulting and books. There is a great incentive to use the phrase and none to make sure that they are.

Michael has researched best practices as part of his PhD dissertation and in their 5-year stint as a management consultant to construction contractors. As a result, he was able to gather much data through clients. Although the data and its statistical analysis will never be shared with anyone else, he has proven to himself what a beneficial technique is valid or, as he terms it, "good operating practice".

The statistical analysis that Michael undertook correlated practices with a firm's overhead/direct cost ratio compared to its peers. This proved to him that many methods were indeterminate and the rest were "predictors". What was important to him is to know what added or did not add to a firm's efficiency, i.e., how much overhead is needed to manage direct cost.

The direct cost to overhead ratio is a compelling metric. If a firm's ratio was higher than its peers, it has been the best chance of making a higher-than-average profit. Additionally, he knew that comparing a net profit number would be misleading since many firms use tax-saving strategies to lower their tax bills. As a result, some numbers do not reflect the robustness of a firm's system and operations, a.k.a. profitability.

There are three current issues in finance and accounting with Dann Electrical: 1) The use of an overwhelming number of cost codes. 2) The format and calculations on job cost report 3) The complicated process of paying supplier invoices 4) The number of separately printed reports.

The previous owner of the construction firm relied heavily on the company's CFO for advice and counsel. Why not? The person is a CPA, originally from a large public accounting firm and has a sharp mind. Therefore, tighter and more invasive accounting processes were enacted over the years. Some were good and some were not in Michael's view. However, their research and client work made it clear that financial and accounting practices have no significant impact on a critical metric: overhead to direct cost ratio – an efficiency metric. Michael believes what is important is that the accounting procedures are accurate and timely.

So, Michael believes that this will not be a quick and radical transformation. Too many people are involved in a complex process. How should he frame his thinking and process?

Questions

1   Should Michael start changing all processes at once or designate one or two and work on those while leaving the others static? Why?
2   Which is a better starting point – changing a person or a process? Why?
3   At the end of the transformation process, is Michael done with changing the firm? Why or why not?

Exercises

1   Create a chart of 6 practices that represent most of the firm's accounting and financial management process.

    1   Accounts Payable.
    2   Accounts Receivable.
    3   Project Budgeting.
    4   Project Cost Report.
    5   Financial Statement.
    6   Monthly Project and Loss Report.

    Prioritise each (first, second, third and so on) to smooth the transition to improved accounting and financial management. Justify your answers. Please consider the integration of these six areas and the training needed as well as others.

2   Please reflect on the need for accurate and timely accounting and financial management in construction contracting. Consider the role of measurement in continued success and future improvement. List three possible reasons for Michael's beginning focus in this area.

10 Summary and Conclusions

We hope we have earned the right to prescribe a system for young constructors to follow at this point in our professional careers. However, we hesitate due to the innumerable variations of the business and craft. Add significant economic shifts such as labour shortages, political interventions or new types of contracts with their growing number of risk-sharing clauses.

However, we can logically limit such a system to the basic drivers of profitable construction contracting in good faith. Indeed, even outsiders to our industry can agree on many basics.

Overall, there is no one solution in construction. Try as we may, it will always take several approaches working in concert to make a project come in on time and cost with safety while keeping a construction contractor profitable. An older contractor has learned many hard-won lessons on keeping their business and projects running smoothly. they will be the first to tell you that there is no singular approach. "It depends" is not a cop-out but a truthful answer.

This is the answer of the contractor who has earned the practical MBA. They have learned "situational management" through hard knocks, lessons learned, scar tissue and generally overcoming several career crises, they have learned "situational management". This is not a theory, but something that the initial facts learned helps us start thinking of a solution but does not determine our answer. The second set of facts (the other shoe dropping) tells us what to do.

A contractor is much like a mountain climber. He is constantly going uphill against gravity. Just because he traversed the last mile well, he cannot hike the next mile leisurely.

The body and mind forget the pain and hardships it has endured but should remember the intensity needed. Not remembering how hard the last days were can present a real problem facing today's trials.

If you are a mountain climber who has just dropped by helicopter to take over an expedition already embarked, this is even more dangerous. You do not emotionally own the journey. You may have received reports but cannot be sure of the truth or the detail.

A mountain climber may have topped the second-highest peak in the world a dozen times along with others. Still, the inexperienced may ask, "But they have never climbed the big one?" It is of no consequence to other mountain climbers. Their body of work – a dozen successful expeditions up the second tallest peak and other lesser climbs is impressive to those who personally know the toll it takes to do even one.

DOI: 10.1201/9781003290643-10

Sportswriters have often claimed the same about legendary players. "But he never won the big one". You will not hear champions say that of those who have never won. They know how fickle a sport and life can be. Their body of work is enough. Those who do not do, but those who only observe do not deserve the same seat at any table with those that do.

Meaning: Consultants and others may be smart intellectually, but not in practical or emotional terms. They look at a contractor's business from afar in a detached way. Their opinion is just one of many. It does not make them right. Those who have the battle experience are the best to lead the discussion. Consultants should contribute. Companies are well served to promote those who have been with the company in the early days and have the practical and emotional experience to keep things real and possible. They will undoubtedly know what the "bad old days" were and what hard work involves.

As an aside, always remember the bad times and distil lessons from them. It is important to "not come back from hell empty-handed". That hard-won (and rare) knowledge will serve you well at some point in your life.

This may be an excellent analogy to share with your young people. Some middle managers may need to hear it too.

We have studied the subject as part of our focus on construction contracting. We are not the experts. They own and operate these firms. If someone is selling software, TQM, recruiting or other services, they may see their offering as the answer to construction's problems. Perhaps these valuable services are part of the answer but, these solutions must be balanced with other needs. Following one of these as a total focus is to ignore the complexities of running a construction firm.

We have an industry that is sick from a thousand cuts. Many little things compose most of our construction ills. We have all searched the world to answer construction's challenges and some of us are still searching. If you find that one great thing that answers all of our challenges, you will earn significant dollars selling it. You should be well compensated; it is rare in our business.

As multiple approaches are the answer, we have categorised them into three main groups. Any profitable construction business is based on three keystones. All else is subordinate:

- Standardising and Document Practices.
- Training to those Practices.
- Improving those Practices.
- Adherence to the those Practices.
- Risk/Reward Curve Position.

Standardising Practices

Good practices facilitate above average net profit percentages. Thoughtful processes drive costs down and keep turnover the same (if not rising). Effective methods keep corporate expenses lower than the competition for a construction company.

We are in a cost side business and the lesser the cost, the more secure the business. Said another way, if these two business numbers are close together, the company may have one destructive event and cost can quickly become more than turnover (a loss).

Cost discipline is not about buying the less expensive item; it is about the least overall cost, including labour productivity and asset utilisation.

They spend several years creating, observing and documenting these practices of better than average contractors give us more than one data point to which to refer. We know many clients have stated that they make the most money consistently following a thoughtful approach.

Our background is in construction and we know the power of an effective process with its integrated practices. Two simple methods can be as powerful as an elegant process. Firstly, simplicity means speed and is easier to monitor. Second, in a complicated and chaotic business, we suggest not to add to the complexity and chaos – an essential point for any manager or executive.

We have shared several dozen practices in this book. The reader will be rewarded by considering them. Each may not be new to you but, these practices will constantly prompt you to think of a better way to practice your profession. Sometimes, we may confirm your current practice's value. Telling you that your process is a standard approach among contractors encourages you to keep using it and not get distracted by the latest fad or criticism.

Formally documenting a firm's construction contracting practices in plain language is the first step to improving them. Caution: we have found that the writing and editing process is tedious and trouble for most construction firms.

Process

The process proposed:

- Document practices one day a week in a straightforward format. See below.
- Spend time – one day a week – working with small subject matter teams on five major practice areas.

    a   Business Management.
    b   Work Acquisition.
    c   Project Operations.
    d   Financial Management.
    e   Human Resources.

The topics, order, team composition and schedule are entirely up to the company management. The sequence and time slots, preferably 2-1/2 hours each group and their practices. Each practice will be articulated for each of the ten prompts. The company may add others:

1 Definition.
2 Compelling business reason.
3 Primary responsible person.
4 Supervisor.
5 Practice steps.
6 Intended outcome.
7 Cardinal sins.
8 Data needed.

9 Form needed.
10 Deadline.

- After each workday, circulate the cumulative document to staff for comments, additions and edits.
- Finalise practices and systematise them as much as possible – computer templates, reporting relationships and practice sequence.

Benefits

- *Improvement:* Having a clear articulation of practices allows for gaps to be identified and filled while better ideas can be inserted at other times.
- *Training Purposes:* New Employees have a good starting place for understanding their practices and their role.
- *On-going mindfulness*: At the end of every formal management and leadership meeting – five minutes might be used to review a single practice to re-emphasise its details and order.
- *Risk Management:* Once all practices are articulated, there is less ambiguity about each procedure if someone assumes a new position.
- *Unity:* The practices document book represents "one version of the truth".
- *Efficiency:* Studies have shown an outsized benefit of higher adherence to the construction contracting industry practices due to its chaotic nature – a 2% increase in efficiency for a 1% gain in timely completion.

Do not underestimate the power of quiet thinking as you try to reach for top-quartile performance, thus financial independence. There is always a better way. It is incredible to us that there are people who think working hard is a physical endeavour. We believe it is both a physical and mental endeavour. An owner or manager makes more money per hour by solving for ROI (such as creating efficacious business processes and better-negotiated outcomes) than their physical activity (such as project meetings or toolbox talks).

Your conscientious employees can help here. They have seen things that work for other companies. Either as an observer or while they worked for them. It does not matter. The process helps all people perform better. However, you cannot rely on good people to perform well if others sync with them. These days, all members of your team need to be dovetailing their actions with each other.

So, the process is critical. The process drives the other two parts of our trinity to profitability in our current environment. Compliance is moot without thoughtful approaches, and your risk/reward curve position is dangerous.

Some processes are tactical (daily) and some are strategic (long term). The daily ones can be documented in a business management manual and the long-term ones written in a strategic plan.

Adherence to Practices

Adherence to practices means high predictability in behaviours. Just as a construction company is well regarded when it scrupulously keeps its promises to clients, so is the payoff when everyone inside a company holds their "promise" to follow the company processes—the faster and more accurate (discipline) the compliance, the lower the cost of business. The right employees affect this key variable positively.

Compliance is critical as the contracting market does not allow modest premiums over competitors. It is a cost side business. That is where higher profits are made.

From working with firms in the construction business, we have observed compliance as high as approximately 90% of all processes. These are highly profitable firms. You may think that this is a surprise that compliance is not over 85% – an "HD". It should not be. Construction firms are the "tail on the dog" in many ways. We work for a funder and user (owner) and are subject to a designer's (Engineer or Architect) vision. These parties tend to have their agenda and a highly productive contractor on their projects is not a top concern. As an example, insistence on a joint formal and thorough pre-job planning process is still rare.

The highest we have observed in a contractor's office is 91% compliance. We have asked some stellar contractors what their compliance is on a company-wide basis – on time and done right the first time – and the answer has come back in the 70% range. We would say these are very thoughtful contractors are the best at what they do. Again, the tail on the dog syndrome.

Good and decent people are critical to any compliance improvement. These are employees who seek to improve past themselves. If their compliance is 80%, they strive for better. It is innate. To find these people makes your job of management much more manageable.

These people care about the company. In many cases, the company is you, the owner or manager. If you are a good and decent person, you will attract and retain good and decent people. Leadership training can only take you so far. Character shows and if a good character is not evident, people can see it on the first day on the job.

Your Risk/Reward Curve Position

This is where the business of construction contracting comes together. Find the sweet spot on that curve and your company are at a place where there is a high reward for the risk you are taking.

The risk/reward curve is inverse to the general business curve. Lessen risk and projects earn a higher ROI. It is odd but true. Increase risk and you will make less money on average. Long term, this is accurate. The effort of managing risk has a positive reward.

Some of which are unknown, i.e., who will next manage this country, what an employee may do on the job site regardless of your policy, an owner's financial situation and the like.

Lower risk construction situations (profit impacts) tend to have higher rewards (ROI). Companies that follow a thoughtful process do move upward in profitability because they lessen or eliminate these profit impacts. It is important to note, some of these risks and thus your position on the curve is chosen by you. These choices are clients, markets, projects, locations, methods, products and many others. Saying "no" or "pass" at times is a good business decision. To be fair, most risk in construction is organic. Some say construction is Latin for "high effort and low reward".

These factors 1) *Processes 2) Compliance 3) Risk/Reward Position* drive profit-ability in construction contracting. What is comforting is that the business can be more quantitatively managed. It will evolve; there are no stagnant and successful construction contracting organisations globally. We do not believe in a bloodless business model that discourages people from creative and entrepreneurial thinking. However, we feel strongly that this approach makes cost, schedule, quality and safety more predictable while driving them in the right direction. As a result, business relationships with clients flourish. No more significant goals exist in con-struction contracting, especially when a founder is approaching retirement age and seeks options for exiting their firm.

References

Abramo, G., D'Angelo, C. & DiCosta, F., 2009. 'University-industry collaboration in Italy: A bibliometric examination', *Technovation*, vol. 29, no. 29, pp. 498–507.

Assbeihat, J., 2018. 'Reasons behind subcontractor's defaults in construction projects', *International Journal of Civil Engineering and Technology*, vol. 9, no. 4, pp. 327–338.

Australia Bureau of Statistics, 2022. 8165.0 Counts of Australian Businesses, including Entries and Exits. June 2017 to June 2021.

Australian Industries, 2015–2020. Australian Bureau of Statistics. Canberra.

Billet, S. & Loosemore, M., 2004. 'Workplace participatory practices: Conceptualizing workplaces as learning environments', *Journal of Workplace Learning*, vol. 16, no. 6, pp. 312–324.

Butcher, K., 2006. 'Learning from text with diagrams: Promoting mental model development with inference generation', *Journal of Educational Psychology*, vol. 98, no. 1, pp. 182–197.

Chalker, M. & Loosemore, M., 2016. 'Trust and productivity in Australian construction projects: A subcontractor perspective', *Engineering, Construction and Architectural Management*, vol. 23, pp. 192–210.

Collins, J., 2011. Good to Great. New York: HarperCollins.

Collins, J. & Hansen, M., 2011. Great by Choice. London: Random House Business Book.

Collins, J. & Porras, J., 1994. Built to Last: Successful Habits of Visionary Companies. New York: Harper Business.

Covey, S., 1999. The 7 Habits of Highly Effective People. Mango Publishing. Coral Gables, FL.

FMI Quarterly (2016). Why Large Contractors Fail – A Fresh Perspective. http://www.fmi quarterly.com/index.php/2016/06/14/why-large-contractors-fail-a-fresh-perspective/ accessed, February 14, 2017.

FMI Quarterly 2017. FMI Corporation Raleigh North Carolina.

Galloway, P., Galloway, P., Nielsen, K. & Dignum, J., 2013. Managing Giga Projects. Reston, Va: American Society of Civil Engineers.

Garrett, H. 2011. Understanding millennials to improve recruiting efficiency. CARLISLE BARRACKS PA: ARMY WAR COLL.

Gladwell, M., 2000. The Tipping Point. New York: Little Brown.

Green, S., 2011. Making Sense of Construction Improvement. West Sussex: John Wiley and Sons.

Gurmu, A. T., Aibinu, A. A. & Chen, T. K., 2016. 'A study of best management practices for enhancing productivity in building projects: Construction methods perspectives', *Construction Economics and Building*, vol. 16, pp. 1–19.

Gurmu, A. T., & Aibinu, A. A. 2017. Construction equipment management practices for improving labor productivity in multistory building construction projects. *Journal of Construction Engineering and Management*, vol. 143.

Harnish, V., 2012. Mastering the Rockefeller Habits. Ashburn, VA: Gazelles.

Hassell, S., Florence, S., & Ettedgui, E., 2009. Summary of federal construction, building, and housing related research & development in FY1999. Santa Monica, Ca: RANDas Science and Technology Policy Institute.

Howell, G. A., Ballard, G. & Tommelein, I., 2011. 'Construction Engineering-Reinvigorating the Discipline'. *Journal of Construction Engineering & Management*, vol. 137, no. 10, 740–744.

Hudson, P., 2001, 'Safety management and safety culture: the long, hard and winding road', Occupational Health and Safety Management Systems, 2001, 3–32.

Isaacson, W., 2014. The Innovators: how a group of hackers, geniuses and geeks created the digital revolution. vol. 1, London: Simon and Schuster UK.

Kuhn, T. & Schlegel, J., 1963. The Structure of Scientific Revolutions, 4, Physics Today.

Laufer, A., Hoffman, E., Russell, J., & Cameron, W., 2015. What Successful Project Managers Do. MIT Sloan Management Review. Spring 2015, pp. 42–52.

Levitt, S. D. & Dubner, Stephen J., 2005. Freakonomics: A Rogue Economist Explores the Hidden Side of Everything. London: Penguin/Allen Lane.

Lewis, M., 2017. The Undoing Project. Penguin Random House UK.

London, K. & Siva, J., 2013. 'A reflexive capability pathway to commercialisation of innovations using an integrated supply chain: a case study of an innovation in the Australian residential sector'. In S. Kajewski, K. Manley & K. Hampson (eds), CIB World Building Congress, Brisbane 2013: Construction and Society, Brisbane, Queensland, Australia: QUT ePrints, pp. 1–14.

Lucas, G., Gratch, J., Cheng, L., & Marsella, S., 2015. 'When the going gets tough: Grit predicts costly perseverance', *Journal of Research in Personality*, vol. 50, no. 2015, pp. 15–22.

Mushonga, E., 2015. A costing system for the construction industry in Southern Africa. Doctoral dissertation. University of South Africa.

Parker, D., Lawrie, M. & Hudson, P., 2006. 'A framework for understanding the development of organisational safety culture', *Safety Science*, vol. 44, no. 6, pp. 551–562.

Peters, T., 1994. Thriving on Chaos. New York, NY: Knopf.

Senge, P., 1990. The Fifth Discipline. New York, NY: Random House.

Stevens Construction Institute 2018. The Construction Contractors' Digest. Archived Material from 2002–2017.

Stevens-Day Construction Institute 2018. The Construction Contractors' Digest. Archived Material 2017–2018.

Stevens, M. & Day, J., 2020. A construction management education focus and process direction: the power of focusing on four outcomes using formative teaching, learning and assessment. In Mostafa, S. & Rahnamayiezekavat, P. (Eds.), Claiming Identity Through Redefined Teaching in Construction Programs. (pp. 26–41). IGI Global. https://doi.org/10.4018/978-1-5225-8452-0.ch002

Tzu, S., 1977. The Art of War. Translated by Thomas Cleary. Boston: Shambhala Dragon.

Yang, H., Yeung, J. F. Y., Chan, A. P. C., Chiang, Y. H. & Chan, D. W. M., 2010. 'A critical review of performance measurement in construction', *Journal of Facilities Management*, vol. 8, no. 4, pp. 269–284.

Zhang, B. D. N. & Rischmoller, L., 2020. 'Design Thinking in Action: A DPR Case Study to Develop a Sustainable Digital Solution for Labor Resource Management', Paper presented to IGLC 28, Berkeley, California, USA.

Index

Note: Page numbers in *italic* indicate figures and in **bold** indicate tables.

Printed in the United States
by Baker & Taylor Publisher Services

Printed in the United States
by Baker & Taylor Publisher Services